# Sex Chromosome Abnormalities and Human Behavior

*AAAS Selected Symposia Series*

Published by Westview Press, Inc.
5500 Central Avenue, Boulder, Colorado

for the

American Association for the Advancement of Science
1333 H Street, N.W., Washington, D.C.

# Sex Chromosome Abnormalities and Human Behavior

## Psychological Studies

*Edited by Daniel B. Berch
and Bruce G. Bender*

Routledge
Taylor & Francis Group

NEW YORK AND LONDON

First published in paperback 2024

First published 1990 by Westview Press

Published 2019 by Routledge
605 Third Avenue, New York, NY 10158

and by Routledge
4 Park Square, Milton Park, Abingdon, Oxon OX14 4RN

*Routledge is an imprint of the Taylor & Francis Group, an informa business*

Library of Congress Cataloging-in-Publication Data
Sex chromosome abnormalities and human behavior : psychological
    studies / edited by Daniel B. Berch and Bruce G. Bender.
    p. cm.—(AAAS selected symposium series ; 112)
    Includes bibliographical references.
    1. Human chromosome abnormalities—Psychological aspects—
Congresses. I. Berch, Daniel B. II. Bender, Bruce G.
III. Series.
RB155.5.S47 1990
616'.042'019—dc20                                    89-38212
                                                         CIP

Publisher's Note
The publisher has gone to great lengths to ensure the quality of this reprint but points out that some imperfections in the original copies may be apparent.

ISBN 13: 978-0-367-28712-2 (hbk)
ISBN 13: 978-0-367-30258-0 (pbk)
ISBN 13: 978-0-429-30587-0 (ebk)

DOI: 10.1201/9780429305870

# ABOUT THE SERIES

The *AAAS Selected Symposia Series* was begun in 1977 to provide a means for more permanently recording and more widely disseminating some of the valuable material which is discussed at the AAAS Annual National Meetings. The volumes in the *Series* are based on symposia held at the Meetings which address topics of current and continuing significance, both within and among the sciences, and in the areas in which science and technology have an impact on public policy. The *Series* format is designed to provide for rapid dissemination of information, so the papers are reproduced directly from camera-ready copy. The papers are organized and edited by the symposium arrangers who then become the editors of the various volumes. Most papers published in the *Series* are original contributions which have not been previously published, although in some cases additional papers from other sources have been added by an editor to provide a more comprehensive view of a particular topic. Symposia may be reports of new research or reviews of established work, particularly work of an interdisciplinary nature, since the AAAS Annual Meetings typically embrace the full range of the sciences and their societal implications.

ARTHUR HERSCHMAN
*Head, Meetings and Publications*
*American Association for*
*the Advancement of Science*

# CONTENTS

# TABLES

# FIGURES

# ACKNOWLEDGMENTS

We are most grateful to the contributors of the chapters in this book who participated in the AAAS symposium on which the volume is based. We would also like to thank the authors who, although not involved in the symposium, generously agreed to contribute additional chapters, thereby providing a more comprehensive coverage of the topic with which this book is concerned. We wish to extend our appreciation to William N. Dember for his help and guidance throughout all phases of this endeavor. Thanks are due also to Carrie Mason-Rogers for her invaluable assistance in the editing of the manuscript. We are especially grateful to Jeannine Barbeau, without whose formatting skills (in addition to typing and editing), persistence, dedication, and continuing encouragement this book could not have been completed. Ms. Nancy Sykes of Personal Assistant, Inc. was instrumental in producing a most attractive camera-ready copy in a limited time period. Finally, we are very appreciative of the support provided toward preparation of this book by the Department of Psychology at the University of Cincinnati and the James and Annette Rosenthal Fund.

*Daniel B. Berch*
*Bruce G. Bender*

*Bruce G. Bender*
*Daniel B. Berch*

# 1 Overview: Psychological Phenotypes and Sex Chromosome Abnormalities

A remarkable breakthrough in the young science of cytogenetics occurred in 1956, when Tjio and Levan demonstrated that the diploid number of human chromosomes was 46, not 48 as previously believed. For the first time, the chromosomes could be seen and examined under a microscope, and the genetic cause of some birth defects identified. The presence of an extra chromosome in the twenty-first pair, trisomy 21, was found to underlie Down syndrome. Other autosomal aneuploidies, including trisomy 13 and trisomy 18, were linked to other identifiable syndromes, with physical malformations and mental retardation invariably present.

Sex chromosome abnormalities (SCA) presented an intriguing contrast to autosomal abnormalities. Individuals with SCA demonstrated neither marked physical stigmata nor as diminished intelligence. Two known syndromes were found to result from SCAs. In 1959, an extra X chromosome (the 47,XXY karyotype) was documented in a man with Klinefelter syndrome (Jacobs & Strong, 1959). That same year, a missing X chromosome (the 45,X karyotype) was documented in a woman with Turner syndrome (Ford, Jones, Polani, De Almedia, & Briggs, 1959). In other cases, SCAs not previously associated with identified syndromes, including 47,XXX and 47,XYY, were documented. Questions of whether and how the presence of SCA affects ability and behavior quickly emerged, and a sizable literature of psychological and psychiatric studies evolved. Misunderstanding has surrounded much of this early behavioral research, and despite many efforts to sift through the conflicting conclusions, myths about SCA have persisted for 25 years.

However, in the past decade, the use of more sophisticated methodologies has yielded a body of more objective data, enabling clearer interpretation of results. A small number of researchers from around the world, represented in the chapters that follow, are responsible for much of this work. This volume is based on a symposium, "Cognitive and Psychosocial Dysfunctions Associated with Sex Chromosome Abnormalities," which was presented at the 1986 Annual Meeting of the American Association for the Advancement of Science. Three chapters are included by SCA investigators who were not present at the AAAS meeting but whose work complements the other chapters and provides a more comprehensive treatment of the subject. The various chapters reflect a rich diversity of scientific approaches, including neuropsychology, behavioral genetics, psychoendocrinology, information processing, and cognitive development. The four major subject groups are examined, including 47,XXY, 47,XYY, 47,XXX, and 45,X. Following the seven reports from individual research groups is a chapter that examines special methodological problems encountered by SCA researchers. The final chapter provides a critical commentary on the previous chapters, discussing the contributions of SCA research to our understanding of genetic influences on behavior.

## INITIAL STUDIES

The remainder of this chapter provides an historical overview of the psychological study of SCA along with the physical and psychological phenotype associated with each karyotype. An appreciation of the history of SCA research is essential to perceiving the gradual changes that have evolved in expectations concerning the behavior of individuals with these conditions.

The suspicion that SCA was associated with behavioral abnormality led to a flurry of chromosome screening studies in mental and penal settings during the early 1960s. The captive populations were easily accessed, and the procedure was relatively simple to perform, consisting of a quick scraping of buccal mucosa followed by confirmatory chromosome analysis in discrepant cases. Over 100 such studies of adult groups were conducted, most in the United States and Europe, and the results were intriguing. The incidence of 47,XXY men and 47,XXX women was four to five times greater in mentally retarded or psychotic groups than their background incidence in the newborn population (Polani, 1977). Among prison populations, an increased representation of SCA men was found, particularly those with 47,XYY (Hook, 1979). In contrast to SCA individuals with excess X chromatin, 45,X women were not found with increased frequency in these groups.

Several chromosome studies of groups of unusual children were also reported. Screening of 48 boys from a juvenile detention center revealed none with SCA (Duffy & Cervenka, 1971). In a larger study of 700 children from a child psychiatric clinic, 11 SCA children were found, reflecting an approximate fivefold increase against the newborn rate (Crandall, Carrel, & Sparkes, 1972). Chromosome screening of 2,606 Swedish children from special education classes resulted in identification of 27 SCA children, again reflecting approximately five times greater representation than found in newborn populations (Eriksson, 1972). Increased representation of SCA children has also been reported among groups of language disordered children (Friedrich, Dalby, Staehelin-Jensen, & Bruun-Petersen, 1982; Garvey & Mutton, 1973; Mutton & Lea, 1980).

Taken together, the studies of chromosomes in abnormal populations established risk factors associated with SCA and began the important process of attempting to understand the underlying causal mechanisms. Hook (1979) proposed three hypotheses that might explain the presence of SCA adults in mental and penal institutions. He immediately rejected the first two; the "associative" hypothesis could not account for the findings on the basis of independent factors such as an increase in adverse economic conditions, while the "social" hypothesis failed to explain maladaptation as a result of physical or external phenotypic correlates, such as tallness. Some available evidence was found to support the third, or "neural" hypothesis, implicating abnormal neurological functioning. In apparent agreement, Polani (1977), following extensive review of the same literature, suggested that SCA may result in brain dysfunction. He further hypothesized that altered brain development may occur because of changes in rate of brain growth, citing evidence that additional chromosomes decrease cell division rate. Polani's (1977) hypothesis, which he supported with evidence of changes in SCA dermal ridge count, has received considerable attention and support (Netley, 1983; Rovet & Netley, 1982).

Although the institutional studies marked an important first step in understanding SCA's influence on human development and behavior, their results were often misinterpreted. The investigators of these studies frequently noted the limited generalizability due primarily to sampling bias (Jacobs, Brunton, Melville, Brittain, & McClemont, 1965). Indeed, the several hundred subjects represented less than 1% of the living SCA population and provided little appreciation for the variability of phenotype or the possibility that some SCA individuals might be relatively normal. However, the thrust of public attention directed at this research led quickly to a series of SCA stereotypes. Women with a 47,XXX karyotype were often seen as psychotic, 47,XXY men as mentally retarded and prone to homosexuality, and 47,XYY men as "super males" in possession of the "murder chromosome." In contrast, 45,X women were seen to have little difficulty in their personal adjustment.

Both Hook (1979) and Polani (1977) warned of the complexity and pitfalls of attempting to generalize from the institution screening studies. Polani (1977) noted the need for:

...special caution in pooling results or in comparing biological parameters where the comparisons are derived from studies and surveys conducted in diverse environmental settings. The complexity of the situation is not surprising when we consider the great variability of individuals which, it is difficult to imagine, would be reduced to a simple stereotype by a chromosome anomaly (p. 93).

Polani also emphasized the need for longitudinal studies of the natural histories of these conditions.

## NEWBORN STUDIES

Prospective studies of SCA began in the 1960s and have provided a clearer picture of the developmental significance of these abnormalities. The three largest studies were conducted in Edinburgh, Scotland (Court Brown, 1969), Denver, Colorado (Robinson & Puck, 1967), and Toronto, Canada (Bell & Corey, 1974). Taken together, these researchers screened chromosomes of 140,000 newborns and identified nearly 200 SCA infants. As a result, the opportunity existed to follow unselected groups of subjects, providing a body of objective information from infancy into adulthood.

These longitudinal studies of SCA children have revealed increased occurrence of developmental, language, and learning deficiencies, and some behavioral problems that differentiate them from their chromosomally normal siblings. However, in contrast to the largely abnormal picture provided by the institution studies, the development of unselected SCA children seemed, in many ways, relatively normal. In addition, two other important findings emerged. First, there is considerable phenotypic variability, even among children with the same karyotype. For example, some children have intelligence quotients above the average range, while others have severe learning disorders. Second, the quality of environment is an important determinant of developmental outcome. The SCA does not act in isolation but in combination with other genetic and environmental influences. SCA children from stable families tend to have developmental skills similar to their chromosomally normal brothers and sisters, while children from dysfunctional families show a much greater increase in psychosocial and learning problems (Bender, Linden, & Robinson, 1987). Additional developmental characteristics of the Denver SCA cohort are discussed in the following chapter.

## INDIVIDUAL SCA KARYOTYPES

The other chapters in this volume include a variety of studies of SCA groups. Behavior, cognition, brain organization, and hormonal variations are examined in detail. To facilitate the reader's understanding of these data, a brief overview of the physical and psychological features associated with each karyotype is included in this introduction.

### 47,XXY

About 1 in 850 males is born with a 47,XXY karyotype. Although birthweight may be slightly reduced, and hypospadias and other mild genital anomalies are occasionally found, the physical appearance of these children is generally unremarkable. After age three, 47,XXY boys tend to be tall, with half of them showing height above the 75th percentile (Robinson, Lubs, Nielsen, & Sorensen, 1979). Motor milestones are often slightly delayed, and sensorimotor integration and motor strength tend to be reduced (Salbenblatt, Meyers, Bender, Linden, & Robinson, 1987). These boys typically demonstrate less speed and coordination than their siblings and are seldom accomplished athletes.

Boys with a 47,XXY karyotype also commonly have small testes. Although incomplete development of secondary sexual characteristics has been associated with this condition, most, in fact, enter puberty normally (Salbenblatt et al., 1985). By mid-puberty, they become hypergonadotropic, testicular growth ceases, and testosterone production decreases. The vast majority of reported cases of adults with 47,XXY have been infertile. They usually demonstrate eunuchoidism, and an undetermined number, possibly one in three, develop gynecomastia. The presence of these features in adults constitutes Klinefelter syndrome, a condition first described in 1942 (Klinefelter, Reifenstein, & Albright, 1942). Gynecomastia can be surgically corrected, and a return to masculine features can be accomplished with testosterone supplementation (Becker, 1972). Few reports are available regarding the use of testosterone in early puberty. A trial of small doses of intra-muscular testosterone (100 mg once every four weeks) over two years in nine boys age 13 years resulted in faster growth in height, pubic hair development, and penile growth than in the 12 control XXY boys. No differences in hormone levels were recorded (or expected) as blood sampling took place one month after the injection of testosterone was given. Behavioral assessment using a self-report questionnaire showed a significant change in the level of anxiety and more negative feelings towards parents (Stewart, Bailey, Netley, Rovet, & Park, 1986).

Full-scale intelligence quotients are, on average, about 10 to 15 points lower among 47,XXY boys than controls. Assuming normal distribution of scores, this would find a slight increase in frequency of mental retardation (i.e., scores below 70), while about 5% have IQs above the average range (Bender & Berch, 1987). Impeded language development remains a consistent finding across prospective studies, affecting at least half of the 47,XXY boys (Robinson et al., 1979). Some investigators have reported lower Verbal than Performance IQs (Stewart et al., 1986; Walzer et al., 1986). Impaired language skills have been commonly documented with specific difficulties in word finding (Bender et al., 1983) and auditory memory (Graham, Bashir, Walzer, Stark, & Gerald, 1981). Associated problems were found in the development of reading skills (Funderburk & Ferjo, 1978; Graham et al., 1981). Bender, Puck, Salbenblatt, and Robinson (1986) demonstrated evidence of a specific language-based dyslexia in seven of fourteen 47,XXY boys. It follows that about two-thirds of 47,XXY boys have required special education assistance in school (Netley, 1986; Stewart, Netley, & Park, 1982), although most are adequately served with part-time intervention while remaining mainstreamed for the majority of academic work. The relationship between cognitive deficits, hemispheric organization, and testosterone levels in 47,XXY boys is discussed in the chapter by Netley.

Antisocial behavior and psychopathology have not been found frequently in unselected samples of 47,XXY boys (Robinson, Lubs, & Bergsma, 1979). Nonetheless, some behaviors identified in the early institution studies have also been seen in the prospective studies. Specifically, 47,XXY boys have been described as less active, less assertive, and more susceptible to stress than controls (Ratcliffe, Bancroft, Axworthy, & McLaren, 1982; Stewart, et al., 1982; Walzer et al., 1978). Reporting on a study of thirty-two 47,XXY boys, the largest unselected sample in the world, Stewart and colleagues (1982) found these subjects to differ from controls in measurements of temperament (47,XXY boys tended to be more quiet, less active, and less sociable) and maladjustment (Stewart et al., 1986).

Because the oldest children in the prospective studies are now in adolescence and early adulthood, information about their adult adaptation is not yet available. The most objective information about adult behavior has been reported by Theilgaard (1981), who screened chromosomes of the tallest 15% of males born in Copenhagen, Denmark between 1944 and 1947. Because 47,XXY men are typically tall, this sampling method presumably identified most of the 47,XXY men present, perhaps omitting some shorter ones resulting in unclear effects upon other sample characteristics. The fourteen 47,XXY men were more submissive and anxious, tended to avoid frequent social contact, had less sexual encounters, and felt less masculine than the group of 52 control males. Theilgaard's more recent investigations with this group are described in Chapter 7.

## 47,XYY

Individuals with a 47,XYY karyotype exhibit some similarities to, as well as differences from, 47,XXY males. The extra Y chromosome is associated with tall stature, but not with other physical characteristics. Unlike 47,XXY boys, they are not hypogonadal or infertile, and demonstrate no more or less secondary sexual development following puberty than 46,XY males. Language and learning difficulties are common. Of greatest interest, concern, and controversy is the question of predisposition to psychiatric disturbance and antisocial behavior in 47,XYY males.

The first case of a 47,XYY male was reported in 1961 (Sandberg, Koepf, Ishihara, & Hauschka), but it was not until 1965 that widespread interest in this karyotype was stimulated by the finding that these individuals are overrepresented in maximum security hospitals (Jacobs et al., 1965). While the fivefold increase in 47,XXY and 47,XXX individuals in institutions may be explained by the downward shift of normally distributed intelligence characteristics (Bender & Berch, 1987), the incidence of 47,XYY men in mental-penal settings may be as much as 20 times their newborn rate, a staggering increase. While approximately 1 in 900 males are born with this condition, they may represent 2% of criminally insane inmates (Hook, 1979). Widespread debate focused on whether the Y chromosome, perhaps mediated by hormones (Meyer-Bahlburg, 1974), was responsible for male aggressiveness, and when present in duplicate drove its possessor to violent excess. In 1968, an Australian jury acquitted a man accused of fatally stabbing an elderly woman following testimony that the man had an extra Y chromosome. Although the extent of influence his SCA had on the jury's decision is undetermined, several other defense attorneys have attempted to use the 47,XYY karyotype as evidence that their client could not be held responsible for his biological drive toward violence. However, the logical weakness was recognized by the legal community and proved generally unsuccessful (Russell & F. H. Bender, 1970). The responsible conclusion that began to emerge was that 47,XYY individuals, like others with SCA, constitute a population at risk for maladaptation, and that environment, particularly family dynamics, exerts considerable influence upon the expression of that risk (Meyer-Bahlburg, 1974).

Prospective studies of 47,XYY newborns have greatly increased our understanding of the developmental influence of this condition. Eventually these studies will help determine which factors contribute to the successful or unsuccessful adaptation of 47,XYY newborns. Reporting on the compiled results of studies of forty-three 47,XYY infants, Robinson, Lubs, Nielsen, and Sorensen (1979) found no evidence of physical or behavioral deviation. More than half of this sample exhibited delayed motor and language milestones, and later experienced learning problems in school (Bender, Puck, Salbenblatt, & Robinson, 1984b). As with the 47,XXY boys, those with 47,XYY tended to have difficulty with language and reading.

Studies of temperament and behavior in this worldwide compilation of prospectively studied 47,XYY children did not produce a clearly defined phenotype of behavior disorder. Both Ratcliffe and Field (1982) and Bender et al. (1984b) noted a tendency toward mild depression, possibly occurring secondary to the frustration and disappointment of longstanding school failure. Reports from the single largest unselected 47,XYY sample in the world indicated that high activity level, negative mood, and temper tantrums were observed more frequently among the twelve 47,XYY subjects than controls at three years of age (Ratcliffe et al., 1982). Behavior ratings four years later indicated that differences between the two groups were not significant, but suggested increased temper tantrums in the 47,XYY boys (Ratcliffe, Murray, & Teague, 1986). Details of these findings along with more recent results are described in Chapter 8. In the only study of 47,XYY men selected from a nondeviant population, Theilgaard (1981) found increases in depressive symptoms, criminal arrest, and aggression toward wives, although these characteristics were not uniformly present among the twelve 47,XYY men. Recent results from Theilgaard's studies of 47,XYY men are discussed in Chapter 7.

## 45,X

In 1938, Henry Turner first formally identified a syndrome comprised of sexual infantilism, short stature, primary amenorrhea, webbed neck, and cubitus valgus (wide carrying angle at the elbow). Wilkins and Fleischmann (1944) subsequently demonstrated that ovarian dysgenesis was another cardinal feature of what later became known as Turner syndrome. An assortment of other physical malformations and clinical manifestations have since been detected in individuals with Turner syndrome, albeit with a relatively wide variation of frequency. These include a shield chest, low hairline, various skeletal abnormalities, lymphedema at birth, coarctation of the aorta, recurrent otitis media, hearing loss, and renal abnormalities. It is of historical interest to note that as early as 1749, Morgnani (cited in Turpin & Lejeune, 1969) provided a clinical description of a 66-year-old woman whose anatomical features correspond to the physical stigmata that characterize Turner syndrome.

It was not until 1959 that the sex chromosome constitution of Turner individuals was determined to be abnormal. Ford and his co-investigators demonstrated that a woman presenting with classic clinical features had a 45,X karyotype, indicating complete loss of the second X chromosome. This outcome in fact was predicted five years earlier by Polani, Hunter, and Lennox (1954). It is now known that approximately 55% of females with Turner syndrome have such a karyotype. The remaining 45% have either partial X monosomy or mosaicism; the latter type consists of two cell lines, one with the second X chromosome missing and the other, normal. The incidence of Turner syndrome is about 1 in 2500 live female births.

Both cognitive and psychosocial characteristics of Turner females have been of interest to researchers over the past 25 years, with the former attracting most of the attention. First reports of intellectual status suggested that Turner syndrome is associated with mental retardation or, at best, dull intelligence (Haddad & Wilkins, 1959; Polani, 1960). Using the Wechsler scales, both Cohen (1962) and Shaffer (1962) subsequently showed that while the Verbal intelligence of Turner women fell within the normal range, their Performance IQ was below average, suggesting a selective deficit in spatial ability. These findings were later replicated by Money (1963), Buckley (1971), and more recently by a number of other investigators (see Chapter 3 for an extensive review of this topic). Furthermore, these results led Garron, Molander, Cronholm, and Lindsten (1973) to argue that a depressed Performance IQ in the presence of an average Verbal IQ can account for the earlier evidence that Turner syndrome was associated with a comparatively low level of general intelligence.

Having confirmed the original findings of reduced nonverbal intelligence in women with Turner syndrome, researchers began administering a variety of other types of psychometric measures in order to discern specific components of this cognitive deficiency or "space-form blindness," as termed by Money and Alexander (1966). Studies revealed impairments in right-left directional discrimination (Alexander, Walker, & Money, 1964), and difficulties with arithmetic tasks (Garron, 1977; Money & Alexander, 1966; Shaffer, 1962). In the late 1970s and early 1980s, investigators borrowed more specialized measurement approaches from other domains, such as neuropsychological techniques (cf. McGlone, 1985; Waber, 1979) and information-processing tasks (Rovet & Netley, 1982), in an attempt to delineate central nervous system dysfunction and mental processing deficiencies underlying the poor spatial performance of Turner females. A more detailed description and review of this work appears in Chapter 3 by Rovet. In addition, in Chapter 5, Nyborg addresses the hypothesis that cognitive dysfunction in Turner females may be mediated by abnormal levels of sex hormones.

Relationships between Turner syndrome and major psychiatric problems have also been examined over the past 25 years, albeit to a much lesser extent than intellectual status. While case reports in the later 1950s and early 1960s suggested occurrence of schizophrenia (Raft, Spencer, & Toomey, 1976; Sabbath, Morris, Menzer-Benaron, & Sturgis, 1961), paranoia (Money & Granoff, 1965) manic-depressive disorder (Hoffenberg & Jackson, 1957), and anorexia nervosa (Halmi & DeBault, 1974), studies of larger samples of women with Turner syndrome failed to yield any consistent patterns or markedly increased frequency of severe psychopathology (Garron & Vander Stoep, 1969; Money & Mittenthal, 1970). Nevertheless, findings from other studies have suggested that females with Turner syndrome may exhibit a personality style characterized by low arousal, high stress tolerance, and immaturity (Baekgaard, Nyborg, & Nielsen, 1978; Money & Mittenthal, 1970; Sonis et al., 1983). Details regarding the evidence for this behavior pattern along with its proposed etiologies are reviewed and critiqued in Chapter 4 by McCauley.

## 47,XXX

In contrast to females with Turner syndrome, those with a 47,XXX karyotype do not manifest a recognizable set of physical characteristics. While they tend to be tall, no other somatic features are identifiable. In light of this finding, it is striking that these females are at greater risk for cognitive and behavioral deviation than any other SCA group with single-chromosome aneuploidy.

The first case of 47,XXX, which included normal intelligence with secondary amenorrhea, was reported by Jacobs et al. (1959). Barr, Sergovich, Carr, and Shaver (1969) reviewed 143 cases of triple X including individuals referred for cytogenetic study because of clinical findings, and others from chromosome screening of several phenotypically normal and abnormal populations in a total of 17 countries. Results from this largely biased survey indicated that about two-thirds of the 47,XXX subjects were physically normal; the physical abnormalities in the remaining group included a heterogeneous list of major and minor findings. Of the 47,XXX adults studied, normal puberty, menses, and reproductive systems were documented for 73%. The remaining women demonstrated varying degrees of ovarian dysfunction, the cause of which is unclear.

It is no surprise that psychiatric disturbance and low IQs were found in the 101 47,XXX women identified in institutions for mentally retarded or psychotic adults. The increased incidence of 47,XXX women in these groups over their newborn background rate led the authors to conclude that this SCA predisposes to mental retardation and mental illness. In addition, the absence of any behavioral abnormality in the five 47,XXX adults found in non-psychiatric clinical settings prompted speculation that some unknown proportion of this population is normal and that the behavioral phenotype is likely variable. Finally, the authors indicated that more definitive information about 47,XXX would arrive with findings from studies of individuals identified at birth.

Combined results of prospective studies involving forty-three 47,XXX infants, first published in 1979 (Robinson, Lubs, Nielsen, & Sorensen, 1979), produced a picture of mild developmental impairment. The condition was determined to occur in about 1 in 1200 newborn females. The majority of infants appeared normal at birth with the exception of a slight increase in two minor anomalies, clinodactyly and epicanthal folds. Head circumference was slightly reduced below the mean, and height was found to increase over controls with age. About half were delayed in both receptive and expressive language, and on average, intelligence was lower than among sibling controls. A report published three years later indicated that 74% of the unselected 47,XXX girls had learning disorders, and that while the subjects evidenced problems of reluctance and difficulty forming interpersonal relationships, findings of atypical personality patterns were not common (Stewart et al., 1982).

The adult development of unselected 47,XXX females will be examined during the next several years as these groups of children are now passing through adolescence. Although consistent patterns of behavioral difficulty were not seen in childhood, their future adaptation remains in question. The findings of increased language and learning problems and decreased intelligence suggest that the level of frustration and self-doubt may be high in this group, and the skills that they possess to compete with peers for educational and vocational success are less than average. In combination with findings of overrepresentation in institutional populations, these results underscore the importance of determining the range of adult adaptation of 47,XXX women.

## OTHER SCA CONDITIONS

The contents of this volume are focused upon the most commonly occurring SCA conditions, including those resulting from a single additional X or Y chromosome or the absence of an X chromosome. Other abnormalities involving the sex chromosomes can occur. These include structural abnormalities of the X chromosome, mosaicism, the presence of two or more extra chromosomes, and fragile X syndrome.

### Structural Abnormalities of the X Chromosome

Structural abnormalities include the presence of an X chromosome in which a short arm or long arm of an X chromosome is missing. An isochromosome is formed when chromosome division is at right angles to the long axis, rather than the normal longitudinal division; the parts reattach to form one large chromosome having the two long arms of the original, and one consisting of the two short arms with no long arms (the latter is rarely seen). Translocations occur when a piece of one chromosome breaks off and attaches to another chromosome, and the reciprocal piece from the other is translocated to the X.

As noted above, females with a short arm deletion (46,XXp-) often demonstrate the Turner syndrome stigmata found in those with complete monosomy. Isochromosome of the long arm, in which the short arms are absent, produces similar results. Where isochromosome of the short-arm or, similarly, long-arm deletions (46,XXq-) occur, streak gonads and infertility are often documented, but short stature and other Turner stigmata may not be. These findings have led to the conclusion that the observable features of Turner syndrome are a product of short-arm deletions, while infertility results from monosomy either of the long arm or the short arm (Ferguson-Smith, 1965).

### Mosaicism

Individuals with SCA may have mosaicism or more than one cell line. For example, an SCA male may have some cells with the normal 46,XY complement and others with 47,XXY. In others, three different cell lines may be present. A variety of forms of X chromosome mosaicism have been reported in girls with Turner syndrome mosaicism, including 45,X/46,XX and 45,X/46,XX/47,XXX. However, when most of the girl's cells contain the normal 46,XX complement, the individual is likely to demonstrate few or no Turner stigmata (Bender, Puck, Salbenblatt, & Robinson, 1984a).

## Two or More Extra Chromosomes

Although rarer than aneuploidy involving a single supernumerary chromosome, two or more extra sex chromosomes have been reported. These 48,XXXY men may have the stigmata of Klinefelter syndrome; cognitive abilities are generally more impaired than in 47,XXY men, although mental retardation is not always present. Men having a 48,XXYY karyotype often present with similar features, plus skeletal and cardiac abnormalities. Males with the 49,XXXXY chromosome complement usually exhibit severely depressed intelligence and numerous physical stigmata, and may be mistaken for Down syndrome patients (Nora & Fraser, 1981).

## Fragile X

Fragile X, or marker X, syndrome is the most common single inherited cause of mental retardation. It occurs in about one in 2000 male births. Since it is X-linked, the syndrome is most pronounced in males, usually consisting of mental retardation, large ears, and large testes, and sometimes including autistic behaviors. Heterozygous fragile X females are less obviously affected, although about one-third of them may demonstrate depressed intelligence. Unlike the SCA conditions described in subsequent chapters, fragile X is not a chromosome aneuploidy. Rather, fragile X involves a constriction found near the distal end of the X chromosome and follows an atypical X-linked pattern of inheritance.

## THE SEARCH FOR SCA PHENOTYPES

The expression of SCA is variable, and some individuals appear relatively unaffected. Is it appropriate, then, to attempt to describe SCA phenotypes? The search for SCA phenotypes is valuable if a distinction is first made between phenotypes, archetypes, and syndromes.

The stigmata of Turner syndrome are usually recognized in infancy. Klinefelter syndrome generally emerges after adolescence. However, most SCA children and adults appear physically normal and do not have syndromes. Neither can they be characterized by archetypes of homogeneous profiles of the sort found in a compendium of birth defects. Their phenotypes are characterized by tendencies and subtle features that occur with increased frequency in SCA groups but not invariably in individuals.

Persons with Down syndrome often resemble each other more than their own family members. This is not the case with most SCA individuals. The terms "genomic repertoire" (Ginsburgh & Laughlin, 1971) and "reaction range" (Gottesman, 1963) capture the concept that the same genotype, under different conditions, can develop into a variety of phenotypes. Individuals with the same SCA may have distinctly different abilities and levels of personal adaptation. Accurate definition of SCA phenotypes must include both descriptions of average tendencies as well as the reaction range of each karyotype.

Psychological studies of SCA have been essential to the difficult task of assessing these subtle and variable phenotypes. The task is not complete, and continued investigation will bring us closer to elucidating when and to what degree SCA alters abilities and behavior. Additionally, scientists will ask how SCA directs its influences. Does the altered chromatin result in changes in the developing brain? Do diminished sex hormone levels interfere with the normal hormonal mediation of brain activities? Does impaired ability to perform specific cognitive functions mediate observed changes in social behavior? How does the environment modulate the genetically-imposed risk factors? All of these questions are addressed in the following chapters in an attempt to define the phenotypes and causal mechanisms associated with SCA.

## REFERENCES

Alexander, D., Walker, H. T., & Money, J. (1964). Studies in direction sense: I. Turner's syndrome. Archives of General Psychiatry, 10, 337-339.

Baekgaard, W., Nyborg, H., & Nielsen, J. (1978). Neuroticism and extroversion in Turner's syndrome. Journal of Abnormal Psychology, 87, 583-586.

Barr, M. L., Sergovich, F. R., Carr, D. H., & Shaver, E. L. (1969). The triplo-X female. The Canadian Medical Association Journal, 101, 247-258.

Becker, K. L. (1972). Clinical and therapeutic experiences with Klinefelter's syndrome. Fertility and Sterility, 23, 568-578.

Bell, A. G., & Corey, P. N. (1974). A sex chromatin and Y body survey of Toronto newborns. Canadian Journal of Genetics and Cytogenetics, 16, 239-250.

Bender, B., & Berch, D. (1987). Sex chromosome abnormalities: Studies of genetic influences on behavior. Integrative Psychiatry, 5, 171-178.

Bender, B., Fry, E., Pennington, B., Puck, M., Salbenblatt, J., & Robinson, A. (1983). Speech and language development in 41 children with sex chromosome anomalies. Pediatrics, 71, 262-267.

14

Bender, B., Linden, M., & Robinson, A. (1987). Environment and developmental risk in children with sex chromosome abnormalities. Journal of the American Academy of Child Psychiatry, 26, 499-503.

Bender, B., Puck, M., Salbenblatt, J., & Robinson, A. (1984a). Cognitive development of unselected girls with complete and partial X monosomy. Pediatrics, 73, 175-182.

Bender, B., Puck, M., Salbenblatt, J., & Robinson, A. (1984b). The development of four unselected 47,XYY boys. Clinical Genetics, 25, 435-445.

Bender, B., Puck, M., Salbenblatt, J., & Robinson, A. (1986). Dyslexia in 47,XXY boys identified at birth. Behavior Genetics, 16, 343-354.

Buckley, F. (1971). Preliminary report on intelligence quotient scores of patients with Turner syndrome: A replication study. British Journal of Psychiatry, 7, 105-127.

Cohen, H. (1962). Psychological test findings in adolescents having ovarian dysgenesis. Psychosomatic Medicine, 24, 249-256.

Court Brown, W. M. (1969). Sex chromosome aneuploidy in man and its frequency, with special reference to mental subnormality and criminal behavior. International Review of Experimental Pathology, 7, 31-97.

Crandall, B. R., Carrel, R. E., & Sparkes, R. S. (1972). Chromosome findings in 700 children referred to a psychiatric clinic. Journal of Pediatrics, 80, 62-68.

Duffy, J. C., & Cervenka, J. (1971). Search for XYY syndrome in psychiatrically disturbed children and adolescent juvenile delinquents. Child Psychiatry and Human Development, 2, 50-53.

Eriksson, B. (1972). Sex chromatin deviations among school children in special classes. Journal of Mental Deficiency Research, 16, 97-102.

Ferguson-Smith, M. (1965). Karyotype-phenotype correlations in gonadal dysgenesis and their bearing on the pathogenesis of malformations. Journal of Medical Genetics, 2, 142-153.

Ford, C. E., Jones, K. W., Polani, P. E., De Almedia, J. C., & Briggs, J. H. (1959). A sex-chromosome anomaly in a case of gonadal dysgenesis (Turner's syndrome). Lancet, 1, 711.

Friedrich, U., Dalby, M., Staehelin-Jensen, T., & Bruun-Petersen, G. (1982). Chromosomal studies of children with developmental language retardation. Developmental Medicine and Child Neurology, 24, 645-652.

Funderburk, S. J., & Ferjo, N. (1978). Clinical observations in Klinefelter (47,XXY) syndrome. Journal of Mental Deficiency Research, 22, 207-212.

Garron, D. C. (1977). Intelligence among persons with Turner's syndrome. Behavior Genetics, 7, 105-127.

Garron, D. C., Molander, L., Cronholm, B., & Lindsten, J. (1973). An explanation of the apparently increased incidence of moderate mental retardation in Turner's syndrome. Behavior Genetics, 3, 37-43.

Garron, D. C., & Vander Stoep, L. R. (1969). Personality and intelligence in Turner's syndrome. Archives of General Psychiatry, 21, 339-346.

Garvey, M., & Mutton, D. E. (1973). Sex chromosome aberrations and speech development. Archives of Disabilities in Childhood, 48, 937-941.

Ginsburg, B., & Laughlin, W. (1971). Race and intelligence, what do we really know? In R. Cancro (Ed.), Intelligence: Genetic and environmental influences (pp. 240-257). New York: Grune and Stratton.

Gottesman, I. I. (1963). Genetic aspects of intelligent behavior. In N. Ellis (Ed.), Handbook of mental deficiency (pp. 182-213). New York: McGraw-Hill.

Graham, J. M., Bashir, A. S., Walzer, S., Stark, R. E., & Gerald, P. S. (1981). Communication skills among unselected XXY boys. Pediatric Research, 15, 562.

Haddad, H. M., & Wilkins, L. (1959). Congenital anomalies associated with gonadal aplasia: Review of 55 cases. Pediatrics, 23, 885-902.

Halmi, K. A., & DeBault, L. E. (1974). Gonosomal aneuploidy in anorexia nervosa. American Journal of Human Genetics, 26, 195-198.

Hoffenberg, R., & Jackson, W. P. (1957). Gonadal dysgenesis: Modern concepts. British Medical Journal, 2, 1457-1462.

Hook, E. B. (1979). Extra sex chromosomes and human behavior: The nature of the evidence regarding XYY, XXY, XXYY, and

XXX genotypes. In H. L. Vallet & I. Y. Porter (Eds.), Genetic aspects of sexual differentiation (pp. 437-463). New York: Academic Press.

Jacobs, P. A., & Strong, J. A. (1959). A case of human intersexuality having a possible XXY sex-determining mechanism. Nature, 183, 302-307.

Jacobs, P. A., Brunton, M., Melville, M. M., Brittain, R. P., & McClemont, W. F. (1965). Aggressive behavior, mental subnormality and the XYY male. Nature, 208, 1351-1354.

Jacobs, P. A., Baikie, A. G., Court Brown, W. M., MacGregor, T. N., MacLean, N., & Harnden, D. G. (1959). Evidence for the existence of the human "super female." Lancet, 2, 423-425.

Klinefelter, H. F., Reifenstein, E. C., & Albright, F. (1942). Syndrome characterized by gynecomastia, aspermatogenesis without A Leydigism, and increased excretion of follicle stimulating hormone. Journal of Clinical Endocrinology, 2, 615-620.

McGlone, J. (1985). Can spatial deficits in Turner's syndrome be explained by focal CNS dysfunction or atypical speech lateralization? Journal of Clinical and Experimental Neuropsychology, 7, 375-394.

Meyer-Bahlburg, H. (1974). Aggression, androgens, and the XYY syndrome. In R. Freedman, R. Richort, & R. Vander Wiele (Eds.), Sex differences in behavior (pp. 433-453). New York: John Wiley and Sons.

Money, J. (1963). Cytogenetic and psychosexual incongruities with a note on space-form blindness. American Journal of Psychiatry, 119, 820-827.

Money, J., & Alexander, D. (1966). Turner syndrome: Further demonstration of the presence of specific congenital deficiencies. Journal of Medical Genetics, 3, 47-48.

Money, J., & Granoff, D. (1965). IQ and the somatic stigmata of Turner's syndrome. American Journal of Mental Deficiency, 70, 69-77.

Money, J., & Mittenthal, S. (1970). Lack of personality pathology in Turner's syndrome: Relation of cytogenetics, hormones and physique. Behavior and Genetics, 1, 43-56.

Mutton, D. E., & Lea, J. (1980). Chromosome studies of children with specific speech and language delay. Developmental Medicine and Child Neurology, 22, 588-594.

Netley, C. (1983). Sex chromosome abnormalities and the development of verbal and nonverbal abilities. In C. Ludlow & J. Cooper (Eds.), Genetic aspects of speech and language disorders (pp. 179-195). New York: Academic Press.

Netley, C. (1986). Summary overview of behavioural development in individuals with neonatally identified X and Y aneuploidy. Birth Defects: Original Article Series, 22, 293-306.

Nora, J. J., & Fraser, F. C. (1981). Medical genetics: Principles and practice. Philadelphia: Lea & Febiger.

Polani, P. E. (1960). Chromosomal factors in certain types of educational subnormality. In P. W. Bowman & H. B. Mautner (Eds.), Mental retardation: Proceedings of the First International Congress (pp. 421-438). New York: Grune & Stratton.

Polani, P. E. (1977). Abnormal sex chromosomes, behaviour and mental disorder. In J. M. Tanner (Ed.), Developments in psychiatric research (pp. 89-128). London: Hoddler and Stughton.

Polani, P. E., Hunter, W. F., & Lennox, B. (1954). Chromosomal sex in Turner's syndrome with coarctation of the aorta. Lancet, 2, 120-121.

Raft, D., Spencer, R. F., & Toomey, T. C. (1976). Ambiguity of gender identity fantasies and aspects of normality and pathology in hypopituitary dwarfism and Turner's syndrome: Three cases. The Journal of Sex Research, 12, 161-172.

Ratcliffe, S. G., Bancroft, J., Axworthy, D., & McLaren, W. (1982). Klinefelter's syndrome in adolescence. Archives of Diseases in Childhood, 57, 6-12.

Ratcliffe, S. G., & Field, M. A. S. (1982). Emotional disorder in XYY children: Four case reports. Journal of Child Psychology and Psychiatry, 23, 401-406.

Ratcliffe, S. G., Murray, L., & Teague, P. (1986). Edinburgh study of growth and development of children with sex chromosome abnormalities. III. Birth Defects: Original Article Series, 22, 73-118.

Ratcliffe, S. G., Tierney, I., Nshaho, J., Smith, L., Springbett, A., & Callan, S. (1982). The Edinburgh study of growth and

18

development of children with sex chromosomal abnormalities. Birth Defects: Original Article Series, 18, 41-60.

Robinson, A., Lubs, H., & Bergsma, D. (Eds.) (1979). Sex chromosome aneuploidy: Prospective studies on children. Birth Defects: Original Article Series, 15.

Robinson, A., Lubs, H., Nielsen, J., & Sorensen, K. (1979). Summary of clinical findings: Profiles of children with 47,XXY, 47,XXX, and 47,XYY karyotypes. Birth Defects: Original Article Series, 15, 261-266.

Robinson, A., & Puck, T. (1967). Studies on chromosomal nondisjunction in man, II. American Journal of Human Genetics, 19, 112-129.

Rovet, J., & Netley, C. (1982). Processing deficits in Turner's syndrome. Developmental Psychology, 18, 77-94.

Rovet, J., & Netley, C. (1983). The triple X syndrome in childhood: Recent empirical findings. Child Development, 54, 831-845.

Russell, D. H., & Bender, F. H., (1970). Legal implications of the XYY syndrome. Seminars in Psychiatry, 2, 40-52.

Sabbath, J. C., Morris, T. A., Menzer-Benaron, D., & Sturgis, S. H. (1961). Psychiatric observations in adolescent girls lacking ovarian function. Psychomatic Medicine, 23, 224-231.

Salbenblatt, J. A., Meyers, D. C., Bender, B. G., Linden, M. G., & Robinson, A. (1987). Gross and fine motor development in 47,XXY and 47,XYY males. Pediatrics, 80, 240-244.

Salbenblatt, J. A., Bender, B. G., Puck, M., Robinson, A., Falman, C., & Winter, J. (1985). Pituitary-gonadal function in Klinefelter syndrome before and during puberty. Pediatric Research, 19, 82-86.

Sandberg, A., Koepf, T., Ishihara, S., & Hauschka, J. (1961). An XYY human male. Lancet, 2, 488-489.

Shaffer, J. W. (1962). A specific cognitive deficit observed in gonadal aplasia (Turner's syndrome). Journal of Clinical Psychology, 18, 403-406.

Sonis, W. A., Levine-Ross, J., Blue, J., Cutler, G. B., Loriaux, P. L., & Klein, R. P. (1983). Hyperactivity and Turner's Syndrome. Paper presented at the meeting of the American Academy of Child Psychiatry, San Francisco.

Stewart, D. A., Bailey, J. D., Netley, C. T., Rovet, J., & Park, E. (1986). Growth and development from early to midadolescence of children with X and Y chromosome aneuploidy. Birth Defects: Original Article Series, 22, 119-182.

Stewart, D. A., Bailey, J. D., Netley, C. T., Rovet, J., Park, E., Cripps, M., & Curtis, J. A. (1982). Growth and development of children with X and Y chromosome aneuploidy from infancy to pubertal age: The Toronto study. Birth Defects: Original Article Series, 18, 99-154.

Stewart, D. A., Netley, C. T., & Park, E. (1982). Summary of clinical findings of children with 47,XXY, 47,XYY and 47,XXX karyotypes. Birth Defects: Original Article Series, 18, 1-5.

Theilgaard, A. (1981). The personalities of XYY and XXY men. In W. Schmid & J. Nielsen (Eds.), Human Behavior and Genetics (pp. 75-84). Amsterdam: Elsevier/North Holland Biomedical Press.

Tjio, J. H., & Levan, A. (1956). The chromosome number in man. Hereditas, 42, 1-6.

Turner, H. H. (1938). A syndrome of infantilism, congenital webbed neck, and cubitus valgus. Endocrinology, 23, 566-574.

Turpin, R., & Lejeune, J. (1969). Human afflictions and chromosomal aberrations. Oxford: Pergamon.

Waber, D. P. (1979). Neuropsychological aspects of Turner's syndrome. Developmental Medicine and Child Neurology, 21, 58-70.

Walzer, S., Bashir, A. S., Graham, J. M., Silbert, A. R., Lange, N. T., DeNapoli, M. F., & Richmond, J. B. (1986). Behavioral development of boys with X chromosome aneuploidy: The impact of reactive style on the educational intervention for learning deficits. Birth Defects: Original Article Series, 22, 1-21.

Walzer, S., Wolff, P. H., Bowen, D., Silbert, A. R., Bashir, A. S., Gerald, P. S., & Richmond, J. B. (1978). A method of the longitudinal study of behavioral development in infants and children: The early development of XXY children. Journal of Child Psychology and Psychiatry, 19, 213-229.

Wilkins, L., & Fleischmann, W. (1944). Ovarian agenesis: Pathology, associated clinical symptoms, and the bearing on the theories of sex differentiation. Journal of Clinical Endocrinology, 4, 357-375.

*Bruce G. Bender*
*Mary Linden*
*Arthur Robinson*

# 2

## SCA: In Search of Developmental Patterns

Abnormalities of chromosome number occur in about one of every 200 newborns. Approximately half of these cases involve sex chromosome abnormalities (SCA). It is known that the effects of SCA are not as severe as those of autosomal abnormalities; for example, most SCA children are not mentally retarded. Still, SCA is widely misunderstood, largely because of the bias introduced by early studies of SCA adults whose phenotypic abnormalities brought them to medical attention.

Prospective studies of SCA newborns identified in screening programs have provided the opportunity to observe the developmental course of these conditions. The Denver SCA study was the first such project in the world. Amniotic membranes of 40,000 consecutive newborns were examined for abnormalities of the sex chromosomes between 1964 and 1974 (Robinson & Puck, 1967). Chromosome analysis of peripheral blood cells was conducted in each case of discrepancy between sex chromatin determinations of the amniotic membrane of the placenta (for Barr bodies), or the transected Wharton jelly of the umbilical cord (for Y bodies), and phenotypic sex of the baby. Because staining techniques for identification of the Y body were developed after X chromosome techniques, only the last 15,641 infants were screened for Y chromosome. Of 68 SCA infants thus identified, 47 became the group followed longitudinally to date and described here (8 infants died in the neonatal period; the remainder declined for various reasons to participate in the full longitudinal study, most because they moved away from Denver).

Table 2.1 indicates who the subjects are: 14 boys with an extra X chromosome; 4 boys with an extra Y chromosome; 11 girls with an extra X chromosome; 9 girls with complete or partial X monosomy (all of whom have varying degrees of Turner syndrome); and 8 girls with sex chromosomal mosaicism. Our controls include 16 males and 16 females who are siblings of various subjects. No more than one control sibling was selected from any participating family. Controls and SCA children have been followed with the same evaluation protocol.

TABLE 2.1
Denver Study Subjects

| Karyotype Subjects | Number of | Mean Age |
|---|---|---|
| 47,XXY | 14 | 14.0 |
| 47,XYY | 4 | 12.5 |
| 47,XXX | 11 | 17.5 |
| 47,X & variants | | |
|     45,X | 6 | 14.9 |
|     46,XXq- | 2 | 17.0 |
|     45,X/46,X,r(X) | 1 | 13.3 |
| Female Mosaics | | |
|     45,X/46,XX/47,XXX | 1 | 20.0 |
|     45,X/46,XX | 4 | 19.9 |
|     46,XX/47,XXX | 1 | 13.0 |
|     45,X/47,XXX | 2 | 17.6 |
| 46,XY Controls | 16 | 14.3 |
| 46,XX Controls | 16 | 14.7 |

The purpose of this chapter is to summarize what we believe to be some of the important findings emerging from this 23-year prospective investigation. Many results have been reported individually. Only now, as most subjects are in adolescence, are we in a position to begin to understand the logitudinal importance of specific developmental findings. SCA children are not all the same. Some have had numerous difficulties that lead us to describe them as "high risk" children; that is, they are poorly equipped to become self-sufficient adults. Others have had relatively normal development. Following are the results of a number of studies that together present a comprehensive picture of the development of these children.

## INTELLIGENCE

Figure 2.1 includes intelligence quotients, or IQs, from the Wechsler Intelligence Scale for Children, administered at nine years of age. All nonmosaic subjects have been grouped together here to provide a global picture of intellectual development. These children have IQs which, on average, are reduced by about one standard deviation relative to controls. Mental retardation is not common, occurring in only three children. There is considerable variability, with IQs ranging from 50 to 122. The mosaics are grouped separately because, unlike the nonmosaic or "pure karyotype" subjects, they do not, as a group, differ from controls. The effects of their abnormal cell lines are apparently offset by their normal cell lines. In this case, their average IQ is similar to that of the controls.

Mean full-scale IQs are presented by individual karyotype group in Figure 2.2. The number of subjects in each group is small, thus limiting our ability to generalize. It can be observed that mean IQs of SCA males, whether having an extra X or Y chromosome, are approximately 10 points lower than those of male controls. The XYY group, consisting of four boys, is too small for statistical comparison to the controls, although their IQ scores appear to be similar to those of the XXY group. The general depression in intelligence is even greater among our female subjects, whether they have an extra X or are missing all or part of an X, with mean scores about 20 points lower than female controls.

## LEARNING, MOTOR, AND LANGUAGE DEVELOPMENT

Figure 2.3 includes results of our investigations of three specific developmental deficits in childhood, including language, motor, and learning disorders. The first involved blind evaluation of SCA children and controls by a speech-language pathologist using standardized instruments (Bender et al., 1983). A similar blind evaluation was conducted by a developmental physical therapist to assess motor skills (Salbenblatt, Meyers, Bender, Linden, & Robinson, 1987). In both studies, the examiners were usually unable to distinguish the appearance of SCA children from that of controls, because with the exception of girls with Turner syndrome, the children do not present with obvious physical abnormalities. The children judged to have language or motor disorders were those whose composite scores fell below the 10th percentile for age. Learning disorders were determined not in the study clinic, but by the actual school adaptation of each child. Those with academic problems requiring part- or full-time special education placement by their schools were included as "learning disordered" (Bender, Puck, Salbenblatt, & Robinson, 1986). All three developmental problems (language, motor, and learning disorders) occurred often in all SCA groups (Figure 2.3).

Language deficits were found most often in XXY boys and XXX girls, while motor deficits occurred with greater frequency among the X monosomy and XXX groups. Learning disorders were identified in 33 (87%) of the nonmosaic SCA children. Males experienced learning disorders more often than females, although their difficulties were primarily in reading and language-related subject areas, and intervention usually consisted of part-time special education. Nonmosaic females often demonstrated more severe deficits across subject areas: 6 of the 20 girls were placed in full-time special education classes. The incidence of all three disorders in the mosaic group was no greater than among controls.

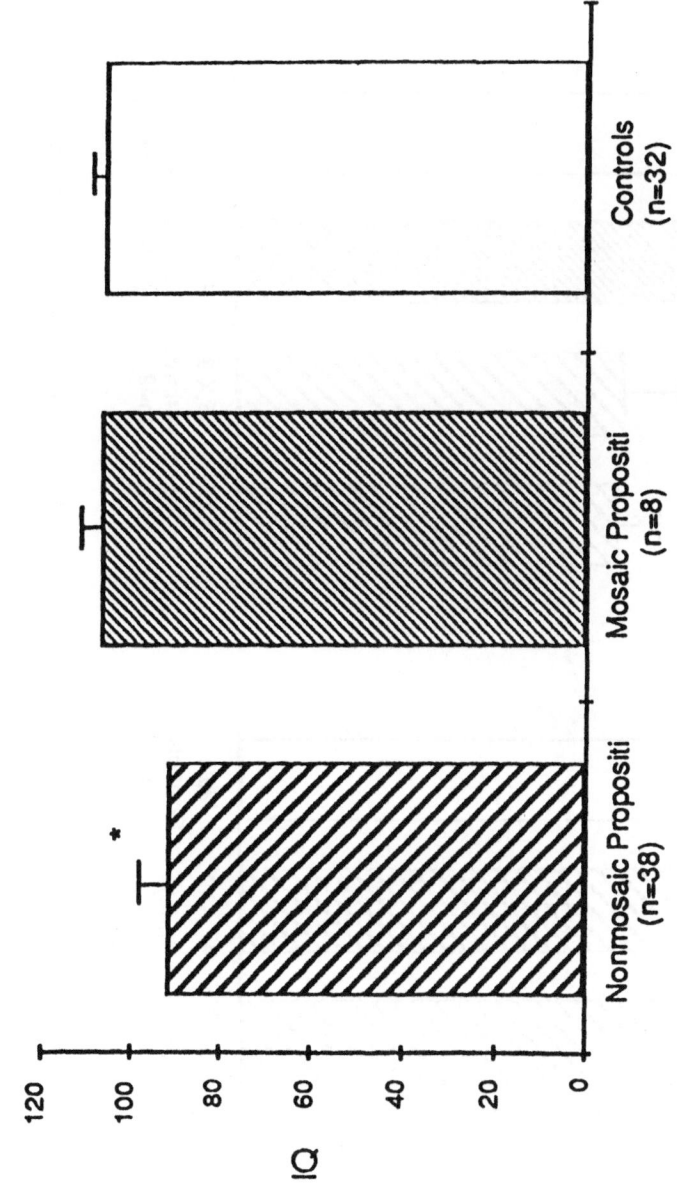

**Figure 2.1** Mean full-scale IQs of propositi (SCA children) and controls at nine years of age. (*Significantly different from controls, $p < .05$.)

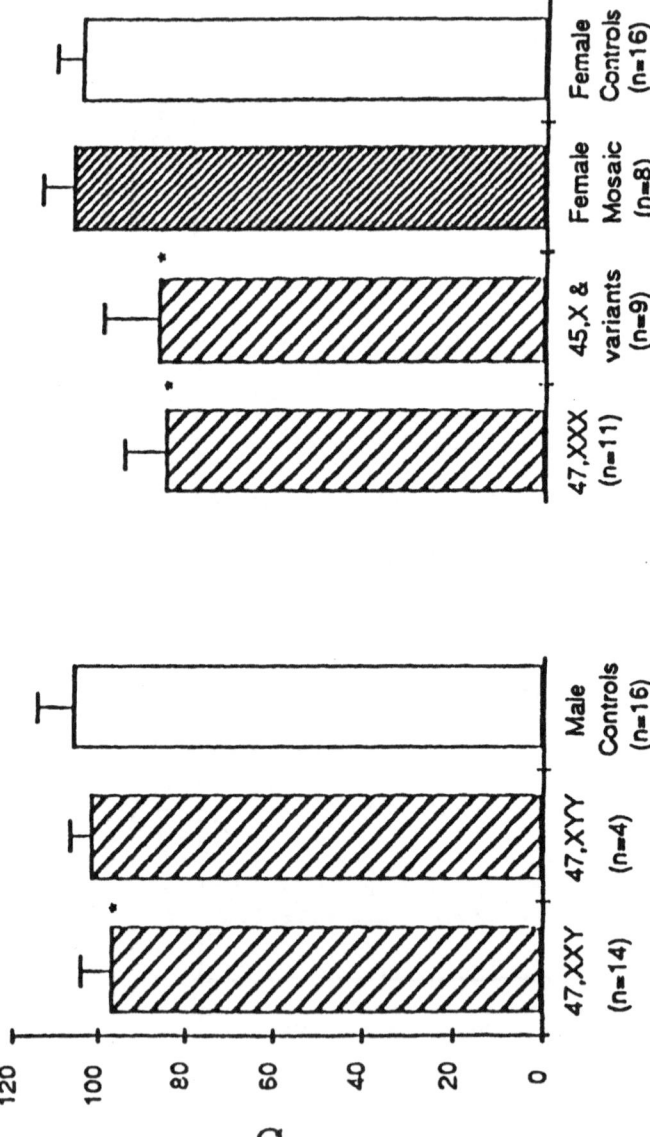

**Figure 2.2** Mean full-scale IQs by karyotype. (*Significantly different from controls, $p < .05$.)

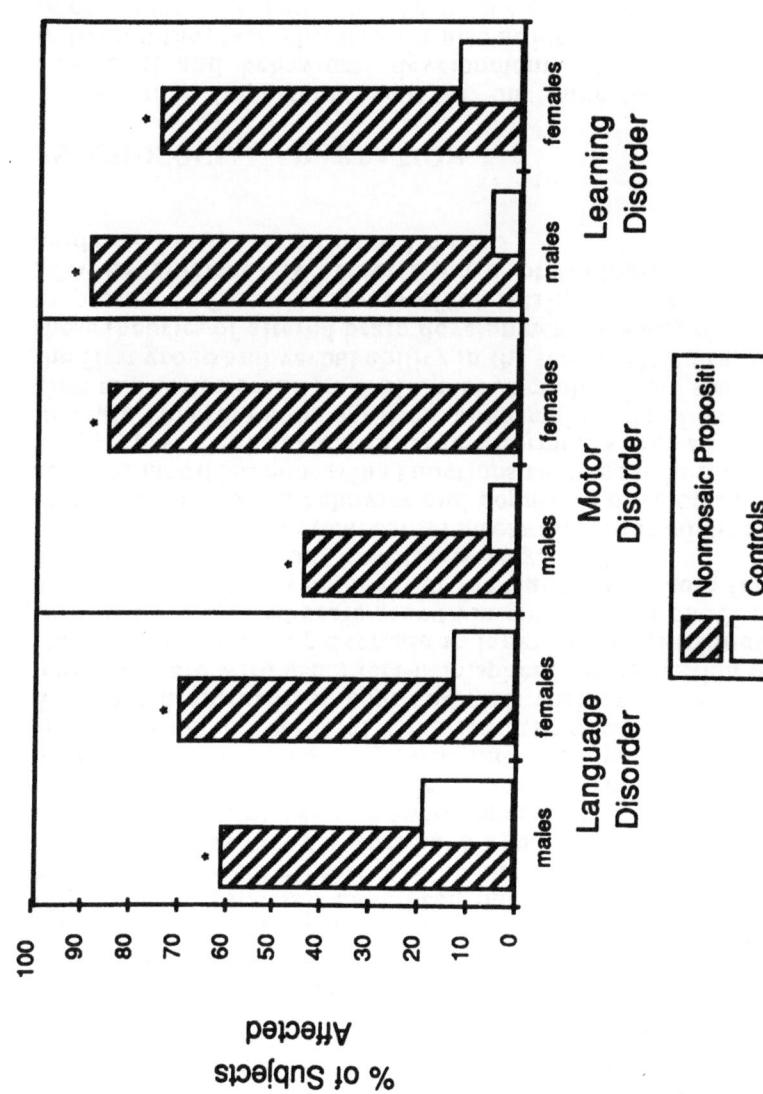

**Figure 2.3** Incidence of specific developmental disorders in nonmosaic propositi (SCA children) and control children. (*Significantly different from controls, *p* < .05.)

These three disorders did not occur independently. Children with language deficits also tended to be those with motor deficits and learning problems. Children with "neurocognitive impairment" were those in whom all three disorders occurred together (Figure 2.4). None of the controls or mosaics exhibited all three deficits. In contrast, 53% of the nonmosaic SCA children had this neurocognitive impairment, which we believe to reflect variations in the development of the central nervous system, affecting both motor and cognitive skills. Again, females were affected more frequently than males. To date, it is not clear how or where the SCA imposes itself on neurological development, although several intriguing possibilities are examined in subsequent chapters in this volume. Two physiological processes have been implicated as potential mediators between the direct effects of SCA and observed behavioral changes. First, numerous investigators have hypothesized that SCA may alter normal patterns of brain growth. Specifically, the presence of SCA is believed to modify cell cycle duration, resulting in increases and decreases in rate of tissue growth and, consequently, maturation and specialization of cerebral functions (Barlow, 1973; Polani, 1977). Slowed brain growth in boys and girls with an extra X chromosome may interfere with usual left-hemisphere specialization of linguistic tasks, with a resulting decrease in language skills (Netley, 1983). In contrast, accelerated brain growth in girls missing one X chromosome may prevent full development of the right hemisphere for processing of spatial information (Rovet & Netley, 1982).

A second hypothetical causal mechanism that might explain some of the alterations in behavior and cognitive development takes into account modified endocrine functions that occur in some of the SCA karyotypes. To the extent that these hormones play an important role in brain function and cognition, abnormal levels of estrogen in 45,X girls and testosterone in 47,XXY boys could affect spatial ability in the first group and verbal ability in the second (Nyborg, 1983). While these theories of altered brain development relating to rate of growth or presence of sex hormones offer intriguing explanations of how SCA development is affected on a physiological level, empirical evidence to support either is limited.

## PSYCHOSOCIAL ADAPTATION

Another important focus of our investigation has been on emotional and behavioral development. The early adult studies indicated that SCA children are also at increased risk of psychosocial impairment. We have studied behavior and personality in several ways (Robinson, Puck, Pennington, Borelli, & Hudson, 1979; Robinson et al., 1982). Although some behavioral difficulties were noted in all nonmosaic groups, significant increases in psychiatric disturbance relative to controls was not found. 47,XXY, 47,XXX, and X

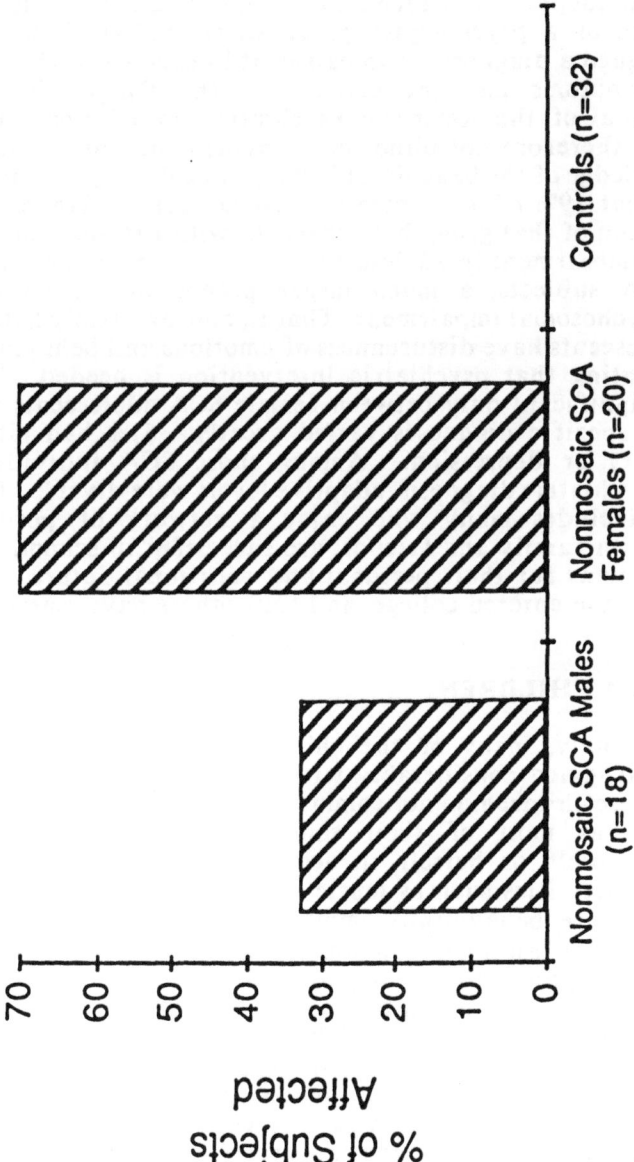

**Figure 2.4** Incidence of neurocognitive impairment (language, motor, and learning disorders) in nonmosaic SCA males and females and in controls.

monosomy children demonstrated a slight but nonsignificant increase in emotional/behavioral disturbance (Robinson et al., 1979).

The data presented here include the most current assessments of the psychosocial adaptation of each adolescent subject. The research team, consisting of a psychologist, psychiatrist, pediatrician, and geneticist, assigned a diagnosis of impairment in each case where an adolescent's symptoms met the criteria of the Diagnostic and Statistical Manual of the American Psychiatric Association. This assessment was therefore not blind, but provided the advantage of extensive knowledge of the behavioral history of each subject. Figure 2.5 indicates that 19% of all controls received such a diagnosis, a sizable proportion of that group but consistent with national surveys of psychiatric impairment in adolescence. Fifty-three percent of all nonmosaic SCA subjects, a much larger proportion, received a diagnosis of psychosocial impairment. That is, approximately half of these SCA adolescents have disturbances of emotions and behavior to a degree indicating that psychiatric intervention is needed. The difference is significant only between female SCA adolescents and controls. The specific diagnoses varied greatly: 61% had either neurotic anxiety or depression. Among those SCA individuals receiving no psychiatric diagnosis, adaptive levels were diverse. This group of 23 includes many who enjoy a network of positive relationships with family and friends, are successful in school, and make realistic plans for their personal and vocational future. For example, three have entered college, and four others have married.

**HIGH RISK SCA CHILDREN**

Not surprisingly, many of the children with neurocognitive deficits have struggled considerably with interpersonal and school adaptation. These children do not communicate as easily or compete as successfully as most of their same-age peers. None has distinguished himself or herself in academic accomplishment or extracurricular activities such as music, social or interest clubs, or athletics. Many are behaviorally immature and socially isolated. Furthermore, school failure is a common experience. Most of the children with neurocognitive impairment have demonstrated psychosocial impairment in adolescence. The group with both impairments includes 16 children, or 42% of all nonmosaics (Figure 2.6). We have labeled this the "high risk group" because it includes those individuals who seem to be least prepared for the challenge of adulthood. The number of "high risk" subjects in each karyotype group is indicated in Table 2.2. Our sample size is small, necessitating caution in generalizing to all SCA individuals. These results indicate again that nonmosaic females are affected much more frequently than nonmosaic males. In addition, girls with an extra X chromosome are represented in greater proportion than X monosomy girls.

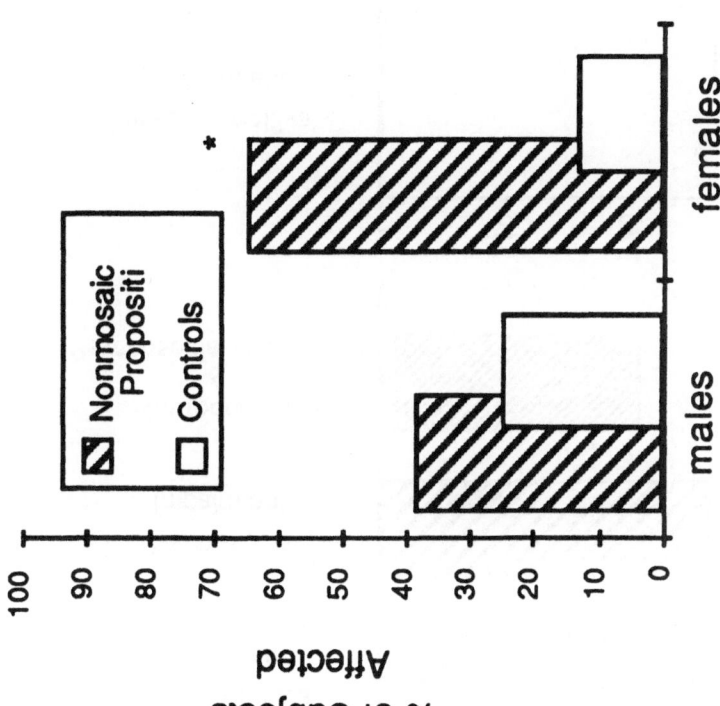

**Figure 2.5** Incidence of adolescent psychosocial impairment in male and female, nonmosaic propositi (SCA children) and controls.

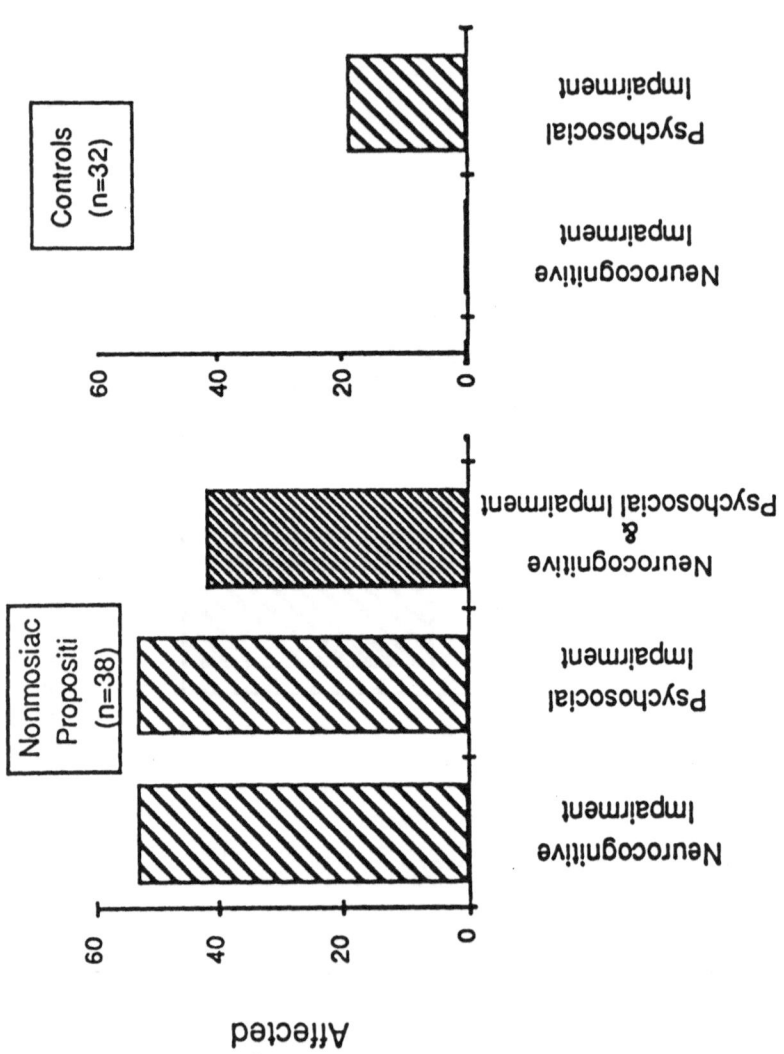

**Figure 2.6  High risk profile:  Neuropsychological impairment in nonmosaic propositi (SCA children).**

TABLE 2.2
High Risk Subjects:  Neurocognitive & Psychosocial Impairment

| 47,XXY | 3 of 14 | | |
| | | 22% of Males | |
| 47,XYY | 1 of 4 | | |
| | | | 42% of Nonmosaic Propositi |
| 47,XXX | 8 of 11 | | |
| | | 60% of Females | |
| 45,X & Variants | 4 of 9 | | |
| Mosaics | 0 of 8 | | |
| Controls | 0 of 32 | | |

What is to happen to these high risk children as they become adults? That is the question for the next several years of investigation. It is not yet clear whether the incidence of SCA psychosocial impairment reported here, which is higher than that noted in earlier reports, represents a temporary dysfunction during a difficult period of adolescence or the onset of serious maladjustment. If the latter is the case, we must determine whether karyotype-specific patterns of psychopathology can be identified in adulthood.


## LOW RISK SCA CHILDREN

Most investigative efforts have focused on the occurrence of abnormal development in SCA children. However, the recognition and study of successful adaptation in SCA individuals is of equal importance.  Variability in SCA development indicates a need to evaluate further the interaction between the SCA and other genetic and environmental factors.

Thirty-seven percent of our sample (14 children) can be included in what we have termed the "low risk group" (Table 2.3).  They have neither neurocognitive nor psychosocial impairments, although some

TABLE 2.3
Low Risk Subjects:  No Neurocognitive or Psychosocial Impairment

| | | | |
|---|---|---|---|
| 47,XXY | 8 of 14 | | |
| | | 50% of Males | |
| 47,XYY | 1 of 4 | | |
| | | | 37% of Nonmosaic Propositi |
| 47,XXX | 2 of 11 | | |
| | | 25% of Females | |
| 45,X & Variants | 3 of 9 | | |
| Mosaics | 7 of 8 | | |
| Controls | 26 of 32 | 81% of Controls | |

have specific language, motor, or learning deficits.  We do not know how successful their adaptation to adulthood will be.  However, some may complete their education, obtain gainful employment, establish a satisfying social network, and marry.  In other words, they could lead lives we would describe as "normal."  Three nonmosaic subjects, one 45,X, one XXY, and one XXX, are now attending college.  We have encountered several other SCA adults who have completed college and, in two cases, graduate school.  Others may follow, although it is our expectation that few SCA individuals will enter professional careers.

**ENVIRONMENT**

Recognition of the presence of "low-risk" SCA individuals leads us to another intriguing question:  What are the "protective factors" that insulate some SCA children from the deleterious effects of their genetic condition?  The answer is undoubtedly complex and not yet well understood, but part of the answer must be environment.  The interaction of environment and genotype is well recognized as the determining formula for all phenotypes.  In the case of SCA children, the effects of environmental stress are more damaging than for chromosomally normal children.  Figure 2.7 shows the dramatic

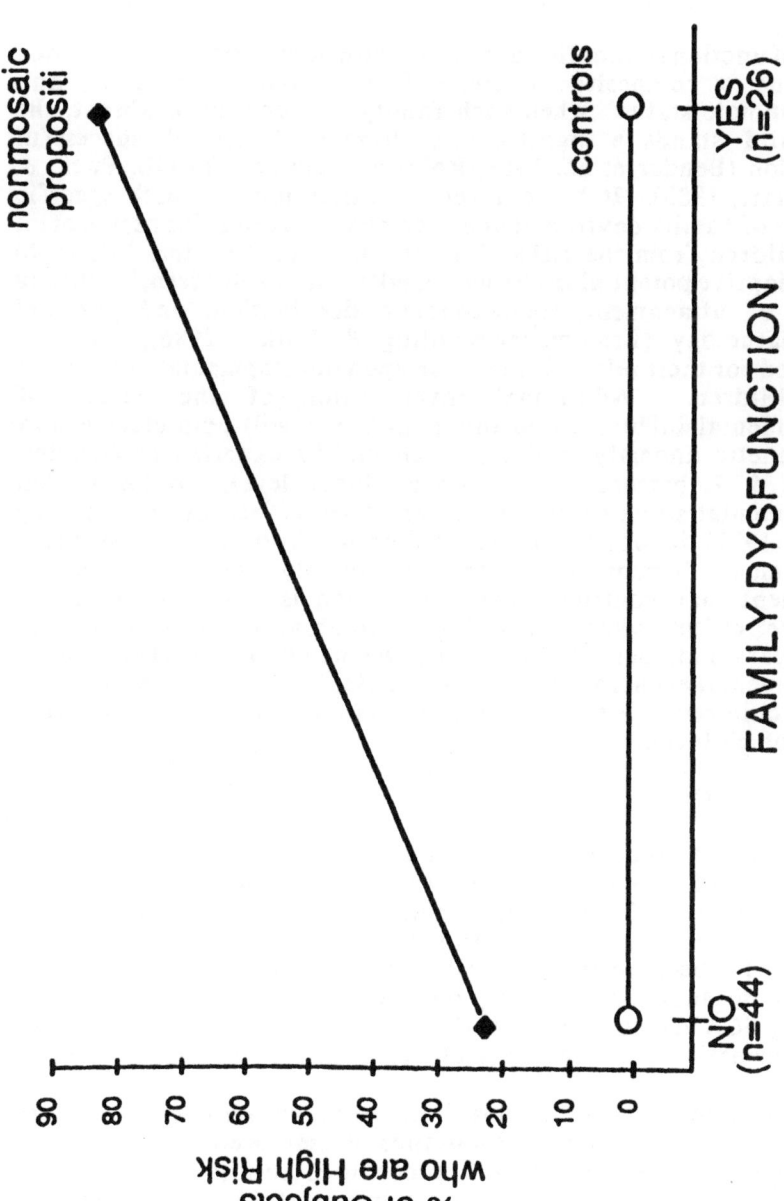

**Figure 2.7** Percentage of high risk nonmosaic propositi (SCA children) and controls from nondysfunctional and dysfunctional families.

increase in "high risk" SCA children emerging from dysfunctional families.

Dysfunctional families are those with low ratings on an index measuring a composite of stress factors, parenting skills, and socioeconomic status. When such family dysfunction is absent, the SCA child stands a significantly greater chance of successful adaptation (Bender et al., 1986; Robinson, Bender, Borelli, Puck, & Salbenblatt, 1983). It has not yet been determined which specific qualities of family environment or personal relationships best protect SCA children from the risks of their condition. Factors known to shape adaptive potential in chromosomally normal children, including quality of attachment, socioeconomic deprivation, and parental psychopathology (Erlenmeyer-Kimling & Miller, 1986), must be evaluated for their relative influence upon developmental outcome in SCA children. Additional investigation of the nature of environmental influences on this population will help clarify how their genetic anomaly mediates each child's experiences (Bender, Linden, & Robinson, 1987). At a global level, we know that environmental stress creates an "interactive" effect, as described by Rutter (1977) in observing that epileptic children from stressful environments demonstrated disproportionately worse psychological adjustment than controls. This interaction is contrasted with the "additive" effect, wherein two factors combine to form the sum of their separate influences. Eventually, we may know whether specific environmental events are crucial to a particular developmental system in SCA children in a manner that differs from their effect on other groups of children.

## IMPLICATIONS

With a birth incidence of one in 400 live newborns, close to 8,000 children with SCA will be born each year in the United States. Given the large number of individuals among us with SCA, it is remarkable that we have only recently identified the effects of these conditions, undoubtedly because the physical and psychological phenotypes of SCA are so subtle. However, the methodological difficulties of studying SCA do not diminish the importance of its investigation. The information provided by such research may be used in several ways.

First, SCA is now diagnosed in utero by amniocentesis with increased frequency. Of the thousands of amniocenteses performed each year, at least one in 250 procedures results in a prenatal diagnosis of SCA (Nance, Dineen, & Brown, 1985). While most of these couples have given careful consideration to their course of action in the event of trisomy 21, few have thought about, or even heard of, an abnormal sex chromosome constitution. In many cases, an immediate and intense search ensues for information about the prognosis. The nature of the information made available to these couples can greatly influence their decision to continue or interrupt the pregnancy. If medical professionals counseling the couple present a largely negative prognosis, and if in particular they are presented with data from the early studies of institutionalized populations, the couple may be more likely to choose an abortion. Some evidence exists that the frequency of abortion following prenatal SCA diagnosis in European countries may be as high as 78% (Ratcliffe, 1986).

In the Denver SCA Study, the objective of genetic counseling following prenatal SCA diagnosis is to present the couple with information that is as current and objective as possible, without implying that either decision, to interrupt or continue, is preferred. Known developmental risks, including language, motor, and learning problems, are described. Phenotypic variability, the importance of environment, and the current unavailability of information about adult adaptation in the studies of unselected SCA individuals are also discussed.

Second, information about developmental risks associated with SCA is also being used to improve understanding of when, where, and to whom intervention can be effectively provided. Family support, language therapy, special education, and individual counseling, if provided early and in the optimal combination, may ameliorate some of the risks of SCA. No definitive program of preventive intervention of SCA children has yet been proposed. Given the observed phenotypic variability, it seems unlikely that predetermined programs of infant stimulation and preschool intervention would prove useful for all SCA children. Robinson et al. (1983) described an approach of "anticipatory guidance" centered upon careful observation and varying degrees of intervention provided as needed.

No program of remediation has been specifically designed for the learning problems of SCA children. If an SCA child has difficulty learning to read, a remedial reading program can be established in almost any public school. Given the frequency of SCA diagnosis, it may be unreasonable to assume that more specific special education assistance can be tailored to these children. However, there is some evidence that SCA children are underserved by their schools and that additional effort must be exerted to provide help beyond referring them for routine special education diagnosis and intervention. Walzer et al. (1986), in a study of 13 unselected 47,XXY boys, found them to have characteristically pliant and quiet "reactive styles," with the result that they were often judged by teachers to be poorly motivated; requests for programs of educational intervention targeting their language deficits were frequently rejected because the boys were viewed as needing only to work harder. Bender et al. (1986) documented significant neuromotor deficits in a group of eleven 47,XXX girls and although severe enough to impair both academic and social adaptation, attributed the absence of any previous recognition of these deficits to diminished value assigned to motor achievements in girls. The difficulties of both 47,XXY boys and 47,XXX girls demonstrate that a better understanding of learning problems of SCA children is needed to increase the specificity and effectiveness of intervention.

Finally, just as inborn errors of metabolism have taught us about normal biochemical and developmental pathways, so SCA may inform us about the role of sex chromosomes and sex hormones in developing physical and psychological systems, about cognitive mediation of behavioral patterns, about organization of the brain's hemispheres, and about environmental modulation of genetic events.

## REFERENCES

Barlow, D. (1973). The influence of inactive chromosomes on human development. Humangenetik, 17, 105-136.

Bender, B., Fry, E., Pennington, B., Puck, M., Salbenblatt, J., & Robinson, A. (1983). Speech and language development in 41 children with sex chromosome anomalies. Pediatrics, 71, 262-267.

Bender, B., Linden, M., & Robinson, A. (1987). Environment and developmental risk in children with sex chromosome abnormalities. Journal of the American Academy of Child Psychiatry, 26, 499-503.

Bender, B., Puck, M., Salbenblatt, J., & Robinson, A. (1986). Cognitive development of children with sex chromosome abnormalities. In S. Smith (Ed.), Genetics and Learning Disabilities, (pp. 175-201). San Diego: College Hill Press.

Erlenmeyer-Kimling, L., & Miller, N. (Eds.). (1986). Lifespan research on the prediction of psychopathology. Hillsdale, New Jersey: Lawrence Erlbaum Associates, Inc.

Nance, M. A., Dineen, M. K., & Brown, J. A. (1985). Fetal chromosome analysis in Virginia: Results and complications of 2,288 cases. Southern Medical Journal, 78, 944-947.

Netley, C. (1983). Sex chromosome abnormalities and the development of verbal and nonverbal abilities. In C. Ludlow & J. Cooper (Eds.), Genetic aspects of speech and language disorders (pp. 2179-195). New York: Academic Press.

Nyborg, H. (1983). Spatial ability in men and women. Advances in Behaviour Research and Therapy, 5, 89-140.

Polani, P. E. (1977). Abnormal sex chromosomes, behaviour and mental disorder. In J. Tanner (Ed.), Developments in psychiatric research (pp. 89-128). London: Hodder and Stoughton.

Ratcliffe, S. (1986). Introduction: Prospective studies on children with sex chromosome aneuploidy. Birth Defects: Original Article Series, 22, xiii-xv.

Robinson, A., Bender, B., Borelli, J., Puck, M., & Salbenblatt, J. (1983). Sex chromosome anomalies: Prospective studies in children. Behavior Genetics, 13, 321-329.

Robinson, A., Bender, B., Borelli, J., Puck, M., Salbenblatt, J., & Webber, M. L. (1982). Sex chromosomal abnormalities (SCA): A prospective and longitudinal study of newborns identified in an unbiased manner. Birth Defects: Original Article Series, 18, 7-39.

Robinson, A., & Puck, T. (1967). Studies on chromosomal nondisjunction in man, II. American Journal of Human Genetics, 19, 112-129.

Robinson, A., Puck, M., Pennington, B., Borelli, J., & Hudson, M. (1979). Abnormalities of the sex chromsome: A prospective study on randomly identified new borns. Birth Defects: Original Article Series, 15, 203-241.

Rovet, J., & Netley, C. (1982). Processing deficits in Turner's syndrome. Developmental Psychology, 18, 77-94.

Rutter, M. (1977). Brain damage syndromes in childhood: Concepts and findings. Journal of Child Psychology and Psychiatry, 18, 1-21.

Salbenblatt, J., A., Meyers, D. C., Bender, B., G., Linden, M. G., & Robinson, A. (1987). Gross and fine motor development in 47,XXY and 47,XYY males. Pediatrics, 80, 240-244.

Walzer, S., Bashir, A., Graham, J., Silbert, A., Lange, N., De Napoli, M., & Richmond, J. (1986). Behavioral development of boys with X chromosome aneuploidy: Impact of reactive style on the educational intervention for learning deficits. Birth Defects: Original Article Series, 22, 1-21.

*Joanne F. Rovet*

# 3 The Cognitive and Neuropsychological Characteristics of Females with Turner Syndrome

This chapter provides a detailed review of the literature on the cognitive and neuropsychological aspects of the Turner syndrome phenotype. The type of impairment associated with the syndrome is described, along with its prevalence, variability of expression, and potential neurological mechanisms. Although it is recognized that deficits in cognitive functioning can also overlap with psychosocial factors associated with the syndrome, information pertaining to the psychosocial aspects will not be covered in this chapter.

## INTELLECTUAL CHARACTERISTICS

It is well known that females with Turner Syndrome (TS) are at risk for cognitive impairment. Early investigators reported that individuals with this condition were prone to mental deficiency (Grumbach, VanWyck, & Wilkins, 1955; Haddad & Wilkins, 1959; Polani, 1961), particularly if they displayed classic physical stigmata of the syndrome. However in 1962, H. Cohen (1962) and Shaffer (1962) both reported in separate papers that these individuals were more likely to have selective cognitive impairments than generalized mental deficiency. Most affected were their nonverbal, visuo-spatial skills. In contrast, their verbal intelligence was unaffected and typically fell in the average range. These findings provided investigators with a potential link between genetic, sex-related, and cognitive characteristics. As a result, considerable interest was generated in studying the TS phenotype.

Since these early reports, numerous studies have been conducted in this area. The findings of 19 published studies are summarized in Table 3.1. They indicate that despite considerable heterogeneity, as a group, TS females perform about one-half to one standard deviation below average and below control groups on the Performance scales of standardized intelligence tests. They do not differ from females with a normal chromosome complement in their Verbal IQ scores, which are in the normal range. (Because the computation of IQ takes into account both the Verbal and Performance scales, overall IQ would

**TABLE 3.1**
**Mean IQ Scores in Studies of Turner Syndrome (TS)**

| Study | TS Sample | | | Country | Controls | Results | |
|---|---|---|---|---|---|---|---|
| | Size | Age(s) | %45X | | | TS | Controls |
| 1. Shaffer (1962) | 20 | 4-31 | 75 | U.S. | — | VIQ = 106.3<br>PIQ = 87.6<br>VP = 18.7 | |
| 2. Money (1963) | 34 | 5-24 | 73.5 | U.S. | — | VIQ = 104.6<br>PIQ = 87.6<br>VP = 17.0 | |
| 3. Money & Alexander (1966) | 16 | 11-24 | 87.5 | U.S. | — | VIQ = 112.0<br>PIQ = 87.5<br>VP = 24.5 | |
| 4. Buckley (1971) | 12 | Not Reported | Not Reported | Britain | — | VIQ = 98.5<br>PIQ = 85.6<br>VP = 12.9 | |
| 5. Kolb & Heaton (1975) | 26 | — | 0 | U.S. | — | VIQ = 100.0<br>PIQ = 71.0<br>VP = 29.0 | |
| 6. Netley (1977) | 14 | 8-20 | Not Reported | Canada | N=18<br>Same age & ability | VIQ = 97.0<br>PIQ = 87.4<br>VP = 9.6 | |
| 7. Silbert, Wolff & Lilienthal (1977) | 13 | 12-22 | 54 | U.S. | N=13 matched for age, SES, race, grade & marital status | TS<C in PIQ $(p<.025)$ | |

**TABLE 3.1**
**Mean IQ Scores in Studies of Turner Syndrome (TS) (continued)**

| | TS Sample | | | | | Results | |
| Study | Size | Age(s) | %45X | Country | Controls | TS | Controls |
| --- | --- | --- | --- | --- | --- | --- | --- |
| 8. Garron (1977) | 67 | 6-31 | 57 | U.S. | N=67 matched for age | VIQ = 99.4<br>PIQ = 88.2<br>VP = 11.2 | VIQ = 104.5<br>PIQ = 103.1<br>VP = 1.4 |
| 9. Waber (1979) | 11 | 12-22 | 45 | U.S. | N=11 from junior colleges and summer camps; matched for age and verbal ability | VIQ = 90.6<br>PIQ = 85.8<br>VP = 4.8<br>TS<Control on FSIQ & VIQ but not PIQ | VIQ = 104.0<br>PIQ = 105.8<br>VP = -1.8 |
| 10. Gordon & Galatzer (1980) | 14 | 16-31 | 28.5 | Israel | 14 outpatients matched for age, education, SES and handedness | VIQ = 99.6<br>PIQ = 88.4<br>VP = 11.2 | |
| 11. Delsmionio, Lis, Saviolo, Rigon, & Tenconi (1981) | 11 | Not Reported | Not Reported | Italy | C1 - 11 same age, grade, SES C2 - 11 14-yr-olds | T = C2 on VIQ<br>T < C1 and C2 on PIQ | |
| 12. Rovet & Netley (1982) | 31 | 11-38 | 67.7 | Canada | 31 from local schools & community colleges; matched for age & VIP | VIQ = 99.2<br>PIQ = 87.0<br>VP = 12.2 | VIQ = 101.3<br>PIQ = 101.6<br>VP = -0.3 |
| 13. Reske-Nielsen, Christensen, & Nielsen (1982) | 1 | 24 | 0 | Denmark | -- | VIQ = 104<br>PIQ = 90<br>VP = 14 | |

**TABLE 3.1**
**Mean IQ Scores in Studies of Turner Syndrome (TS) (continued)**

| | | | TS Sample | | | Results | |
| Study | Size | Age(s) | %45X | Country | Controls | TS | Controls |
|---|---|---|---|---|---|---|---|
| 14. Pennington, Heaton, Karzmark, Pendleton, Lehman, & Shucard (1985) | 10 | 15-33 | 100 | U.S. | Gp1=20 Normal age<br>Gp2=12 Right hemisphere<br>Gp3=10 Left hemisphere<br>Gp4=10 Diffuse lesions | VIQ = 100.5<br>PIQ = 92.5<br>VP = 7.5 | |
| 15. McGlone (1985) | 11 | 13-18 | 73 | Canada | 22 normal females matched 2/TS for age, education, & hand preference | VIQ = 106.5<br>PIQ = 94.7<br>VP = 11.8 | |
| 16. Lewandowski, Costenbader, Richman (1985) | 10 | 9-33 | 90 | U.S. | 10 matched for age, SES, ethnicity, & hand preference | VIQ = 110.0<br>PIQ = 89.2<br>VP = 20.8 | VIQ = 115.4<br>PIQ = 106.4<br>VP = 9.0 |
| 17. Berch, Kirkendall, Briscoe, Digman, & Smith (1985) | 10 | 8.3 | 100 | U.S. | 10 matched for CA & VIQ | VIQ = 102.4<br>PIQ = 93.2<br>VP = 9.2 | VIQ = 102.2<br>PIQ = 104.9<br>VP = -2.7 |
| 18. Robinson, Bender, Borelli, Puck, Salbenblatt, & Winter (1986) | 8 | 11-18 | 75 | U.S. | Siblings in entire study | VIQ = 90.6<br>PIQ = 85.8<br>VP = 4.8 | VIQ = 104.0<br>PIQ = 105.8<br>VP = -1.8 |
| 19. McCauley, Kay, Ito, Treder (1987) | 17 | 9-17 | 58 | U.S. | 16 short stature, comparable age, VIQ, SES | VIQ = 95.4<br>PIQ = 91.4<br>VP = 4.0 | VIQ = 99.9<br>PIQ = 107.5<br>VP = -7.6 |

generally be lower than normal, falsely giving rise to the impression of moderate mental deficiency [Garron, Molander, Cronholm, & Lindsten, 1973].) Examination of the studies in Table 3.1 reveals that the range of difference between Verbal and Performance IQ scales varies from 4.0 (McCauley et al., 1987) to 24.5 (Money & Alexander, 1966).

Table 3.1 also provides demographic and genetic characteristics of the samples. There appears to be no relationship between the size of the Verbal-Performance discrepancy and either the age, genetic karyotype, or nationality of the different samples of individuals with Turner syndrome.

Table 3.2 shows the mean Verbal IQ (VIQ), Performance IQ (PIQ) and Verbal-Performance (VP) discrepancy for TS and control samples obtained by averaging across the results from 13 of the 19 studies in Table 3.1.[1] Although TS individuals obtained VIQ scores comparable to controls, they differed by about 15 points in PIQ. This analysis of the published research shows that the relative impairment in nonverbal visuo-spatial processing among TS individuals is robust and pervasive.

However, results within the various studies appear to be far less consistent, showing considerable variability among the TS females sampled. For example, Pennington et al. (1985) found that while half of their sample exhibited substantial VIQ-PIQ discrepancies, the remainder had minimal or nonsignificant discrepancies.

TABLE 3.2
Mean IQ Scores of Subjects from 13 Studies in Table 3.1

|  | Turner Syndrome (N = 226) | Control (N = 142) |
| --- | --- | --- |
| VIQ | 100.5 | 104.0 |
| PIQ | 88.7 | 103.8 |
| VP | 11.8 | 0.2 |

## Intelligence and Somatic Stigmata

It is well known that the phenotype associated with TS is quite variable. Some individuals have a number of the physical stigmata characteristic of TS while others have only a few. Simpson (1975) reported the following incidence for some of the more common stigmata: primary amenorrhea (97%), short stature (95%), cubitus valgus (54%), pigmented nevi (63%), short broad necks (74%), shield chests (53%), cardiac abnormalities (13%).

In the published literature, two studies have specifically addressed the issue of the correlation between intelligence and somatic characteristics. In the Money and Granoff (1965) study, information was recorded on 12 somatic anomalies from 44 patients, 32 (73%) of whom had a 45,X karyotype. The mean number of anomalies of 4.52 (range = 1 to 11) was higher in the 45,X group (mean = 5.03) than the 45,X/46,XX mosaics or those with translocation or deletion karyotypes (mean = 3.08). The authors found no significant relationship between the number or type of physical anomaly and IQ or VIQ-PIQ discrepancy. Similarly, in a study of one of the largest TS samples (Garron, 1977), IQ was not associated with any of the six most frequent somatic stigmata (cubitus valgus, neck webbing, shield chests, pigmented moles, digital defects, and cardiac abnormalities). However, individuals with three or more stigmata had lower VIQ and PIQ scores overall than individuals with fewer stigmata. According to Pennington et al. (1985), brain manifestations of Turner syndrome may be just as variable as those for heart, kidney, and other organ systems, which may or may not be affected by the condition.

## Intelligence and Turner Syndrome in Twins

One of the major limitations of the studies summarized in Table 3.1 has been the lack or inadequacy of controls, particularly in those studies involving small samples. The designs used in these investigations have matched for age (Garron, 1978; McGlone, 1985; Pennington et al., 1985), FSIQ (Silbert et al., 1977), Verbal IQ (Berch et al., 1985; McCauley et al., 1987; Rovet & Netley, 1982; Waber, 1979), socioeconomic status, race and/or ethnicity (Dellantonio et al., 1981; Gordon & Galatzer, 1980; Lewandowski et al., 1985), and growth retardation (McCauley et al., 1987; Nielsen et al., 1977). One study has used sisters, but these obviously differ in age (Nielsen et al., 1977). Each method is less than ideal because the full range of variables has not been entirely accounted for in any one study. The optimal approach is to examine twin sisters, wherein one member of the pair has Turner syndrome.

We have previously reported on one twin pair involving a girl with TS and a sister with a 46,XX complement (Rovet & Netley, 1982). The IQ data for this and two additional twin pairs seen more recently are presented in Table 3.3. Unfortunately, information on their zygosity is not available. The results show considerable variability in actual pattern of scores and size of the VIQ-PIQ split among the TS children. Nevertheless, the VIQ-PIQ discrepancy scores were consistently positive for the TS girls and consistently negative for their non-TS twin sisters. When we compared each girl with her non-TS twin, we found that the TS twin obtained significantly lower PIQ scores ($p < .01$). Although the TS twin averaged about a 10-point discrepancy between her own Verbal and Performance scales, she averaged about a 20-point difference when her discrepancy score was compared with that of her non-TS twin sister. Therefore, these results show that not only does the impairment exist, but it may be larger than previously believed.

**TABLE 3.3**
**IQ Scores of TS/Non-TS Twin Pairs**

| | | VIQ | | PIQ | | VIQ-PIQ | | TS Minus Non-TS |
|---|---|---|---|---|---|---|---|---|
| Twin Pair | Age | TS | Non-TS | TS | Non-TS | TS | Non-TS | TS Diff |
| A. | 6 | 103 | 113 | 91 | 128 | 12 | -15 | 27 |
| B. | 12 | 94 | 97 | 80 | 100 | 14 | -3 | 17 |
| C. | 5 | 109 | 112 | 104 | 129 | 5 | -17 | 22 |

## Specific Factor Deficits

The studies indicate consistently that the cognitive deficits seen in TS individuals reflect their poorer performance in two areas of intellectual functioning: spatial ability and numerical skills. They have been shown to score significantly lower on two of J. Cohen's (1957) factors, Perceptual Organization and Freedom from Distractibility, while their Verbal Comprehension factor scores are usually average (Buckley, 1971; Garron, 1977; Money, 1963, 1964; Shaffer, 1962; Theilgaard, 1972). The only exception to these findings in the literature is a study by Waber (1979), in which she reported that the matched controls scored as low as TS subjects on the Perceptual Organization factor.

The Freedom from Distractibility factor scores are composed of the results of three subtests: Arithmetic, Digit Span, and Coding/Digit Symbol. The poorer performance by TS individuals on these tasks typically has been attributed to poor numerical facility or a mild degree of dyscalculia, which is reflected in their difficulty with mathematics at school. However, these results may also be attributed to lower scores on this scale (Kaufman, 1979).

Table 3.4 summarizes our own findings (Rovet & Netley, 1982) on 23 TS subjects and 23 matched normal controls on the three J. Cohen (1957) factor scores. As observed by others, the TS subjects scored significantly lower than controls on both Perceptual Organization and Freedom from Distractibility.

**TABLE 3.4**
**Mean Intelligence Test Factor Scores of Turner Syndrome and Control Subjects**

|  | Turner Syndrome (N=226) | Control (N=142) | Significance Level |
|---|---|---|---|
| Verbal Comprehension | 9.3 | 9.8 |  |
| Perceptual Organization | 7.7 | 10.3 | .002 |
| Freedom from Distractibility | 7.6 | 10.1 | .002 |

## Subtest Profiles

Although the VIQ-PIQ discrepancy consistently reported among TS females has been attributed generally to their poor ability in the nonverbal performance domain, there is considerable variability in the studies when the Wechsler subtests are examined individually. Because subtest profiles often provide some indication as to the type of deficit, it is important to examine closely the actual pattern of scores obtained by TS individuals.

In our study (Rovet & Netley, 1982), we found that the TS subjects differed from the controls in all of the Performance subtests, as well as in Arithmetic and Digit Span. Figure 3.1 shows a relatively marked weakness of TS girls in all subtests of the Performance scale. Nevertheless, the profile patterns were virtually identical for both groups, except for the Block Design subtest. The difference for Block Design, while still significant, was somewhat smaller. These findings suggest a widespread deficiency in all processes characterizing the Performance scale, particularly visual perceptual, visuo-motor, and visual reasoning abilities (Kaufman, 1979). Spatial visualization and/or visual concept formation abilities, as tapped by the Block Design subtest (Kaufman, 1979), appear to be slightly less compromised among these individuals. Regarding the verbal subtests, TS girls obtained almost identical scores in verbal reasoning, conceptualization, and comprehension tasks. Their moderately lower scores on Information and significantly lower scores on Arithmetic and Digit Span may reflect difficulties in memory and/or numerical facility. Poorer performance on these tasks has also been associated with distractibility and the inability to concentrate (Banantyne, 1974).

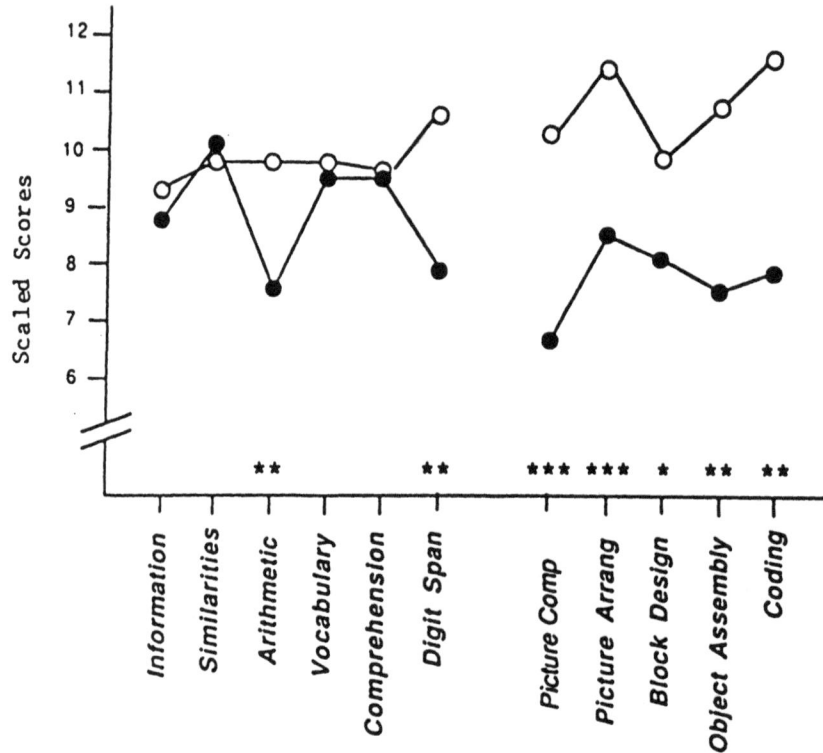

**Figure 3.1** Mean Wechsler subtest scaled scores for 23 Turner (filled circles) and 23 control (open circles) subjects.

Our findings are similar to those reported by Garron (1977) using a much larger TS sample (N = 67). However, recent studies by Waber (1979), Silbert et al. (1977), and McGlone (1985) involving considerably smaller samples but much more detailed assessment procedures and careful matching, have indicated that the deficit is not as extensive as originally believed or as found by Garron (1977) and ourselves (Rovet & Netley, 1982). These researchers found that fewer of the subtests were affected. Waber (1979) reported differences from controls on only the Digit Span task; Silbert et al. (1977) and McGlone (1985) found differences for Picture Completion and Digit Symbol subtests, as well as Digit Span, but not for Block

Design or Object Assembly, the two most "spatial" of the Performance subtests. McGlone also reported that the difference on Digit Span was more pronounced for Digits Backward, which involves the manipulation of numbers as well as recall, than for Digits Forward. This can be interpreted as indicating that the ability to mentally transform (i.e., reverse) numerical symbols in memory may be more impaired than numerical recall. However, because the three studies involved relatively few TS subjects (range = 10 to 13 cases), the likelihood of a Type II statistical error (i.e., reporting no effect when there really is one) was quite large.

Taken together, the results of the studies reviewed here indicate that while the PIQ scores of TS individuals are lower than average, there is neither a homogeneous TS cognitive profile, nor a common intellectual deficit. Further investigations involving larger samples and detailed analyses of subtest profiles are definitely warranted, along with error analyses to provide additional information on the underlying contributing problem.

## Specific Cognitive Deficits

The large Verbal-Performance discrepancy observed consistently on intelligence tests among TS subjects has usually been attributed to a deficiency in the nonverbal visuo-spatial domain of cognitive processing. Of recent concern is whether this indicates a true spatial disability or whether it reflects an associated cognitive problem. For example, the Performance subtests are timed and also have a greater motor component. The above analysis of subtest profiles failed to reveal a common deficit among the TS subjects studied. Furthermore, studies involving more extensive and elaborate testing of TS individuals have shown that certain aspects of verbal processing also appear to be compromised. Table 3.5 summarizes the findings of 20 published studies of TS individuals assessed with procedures that provide information as to specific cognitive impairments.

Examination of the two right-hand columns of Table 3.5 reveals that although a wide range of visuo-spatial skills are affected by TS, other skills such as verbal fluency, auditory sequencing, flexibility, verbal memory, and motor coordination also appear to be problematic. In earlier investigations, visuo-spatial impairments were indeed most common. However, these studies used only visuo-spatial tasks and lacked suitable controls. One exception was a study by Money and Alexander (1966), using the SRA Primary Mental Abilities test, which indicated a deficiency in verbal fluency. Verbal reasoning ability was not affected.

Among the more recent studies, only Waber's (1979) did not show a deficit in the visuo-spatial area. However, this may have been due to sampling bias, including bias in the control group, given that only nine TS females were studied, and two had endogenous hormone production; the controls also showed poorer Block Design performance. Studies using small samples (e.g., McGlone, 1985;

**TABLE 3.5**
**Deficits Reported in Studies of Turner Syndrome Subjects**

| | Study | N | %45X | Age(s) | Controls | Tests Given | Results | Deficits |
|---|---|---|---|---|---|---|---|---|
| 1. | Shaffer (1962) | 20 | 75 | 4-31 | No | Benton Visual Retention | 15/17 below expectancy | Visual memory |
| 2. | Alexander, Walker, & Money (1964) | 13 | Not Reported | 13-26 | Yes | Money Road Map | TS < Controls | Directional sense |
| 3. | Money & Alexander (1966) | 16 | 87 | 11-24 | No | SRA Primary Mental Abilities | 14/16 below 25% in SpatialReasoning 7/16 below in Number 10/16 below in Word fluency Fluency | Space-Form perception; numerical facility; verbal |
| 4. | Alexander & Money(1966) | 16 | 87 | 10-24 | Yes | Benton R-L Disorientation; Detroit Test of Learning Bender Aptitude | TS < C in recognition of R-L on person facing one-self, in imaging rotation of body in space, and in visual copying | Dysgnosia for extrapersonal space and form sperception |
| 5. | Anderson (1968) (cited in Theilgaard, 1972) | -- | -- | -- | No | Nonverbal Memory Test | Deficit in visual perception or Visuo-motor functioning; not visual memory | Visual Perception Visuo-motor functioning |
| 6. | Kolb & Heaton (1975) | 1 | 0 | 24 | No | Halstead-Wepman Aphasia Battery | No language deficit, good auditory perception, difficulty in integrating spatial relations in copying, mild finger dysgnosia | Space-form perception |

**TABLE 3.5**
Deficits Reported in Studies of Turner Syndrome Subjects (continued)

| Study | N | %45X | Age(s) | Controls | Tests Given | Results | Deficits |
|---|---|---|---|---|---|---|---|
| 7. Waber (1976) | 9 | 22 | 13-23 | Yes | Wisconsin Card Sorting Word Fluency Stroop Money Directional Sense CTMM Spatial Ability | TS < C 1-3 but not 4 and 5 | Fluency, flexibility and planning; NOT Spatial Ability |
| 8. Silbert, Wolff, & Lilienthal (1977) | 13 | 54 | 12-22 | Yes | 5-Hour Extensive Battery | TS selective impairment in organizing elements into wholes; perception, visualization, and remembering spatial configurations; TS adequate on pattern analysis (whole into parts); TS have difficulty with temporal analysis of perceptual events difficulty in auditory sequencing | Part-whole perception; Spatial perception; Temporal Analysis; Auditory sequencing |
| 9. Nielsen, Nyborg, & Dahl (1977) | 45 | 47 | 7-39 | Yes | Rod & Frame Test Embedded Figures Money Road Map Porteus Mazes | 47% TS extreme field dependence versus 11% sisters & 27% controls; TS slower at embedded figures; TS poorer on Money Road Map; No difference on Mazes | Field dependence; Spatial ability; Directional sense |

TABLE 3.5
Deficits Reported in Studies of Turner Syndrome Subjects (continued)

| | Study | N | %45X | Age(s) | Controls | Tests Given | Results | Deficits |
|---|---|---|---|---|---|---|---|---|
| 10. | Serra, Pizzamiglio, Boari, Spera (1978) | 9 | 32 | 15-32 | No? | Right-left discrimination Verbal ability | TS Worse on field dependence spatial organization & orientation; not spatial visualization R-L discrimination or verbal ability | Field dependence; Spatial organization; Spatial visualization, |
| 11. | Waber (1979) | 11 | 45 | 12-23 | Yes | Finger Tapping, Stroop CTMM Spatial Face Recognition RL Orientation Consonant Trigrams Word Fluency Money Road Map Key Osterreith Rhythmic Sequencing Card Sorting | TS Poorer on Stroop, visual memory (not specific to nonverbal materials), facial recognition (upright only), word fluency, Road Map, Key Osterreith, card sorting | fluency Planning; Visual memory; Directional sense; Design Copying |
| 12. | Gordon & Galatzer (1980) | 14 | 27 | 16-31 | No? | Manual Dexterity Sequencing Vocabulary | Normal sequential and verbal performance; low Visuo-spatial performance | Visuo-spatial |
| 13. | Rovet & Netley (1982) | 31 | 68 | 11-28 | Yes | Mental Rotation Sentence Verification | TS significantly poorer in rotation task; no difference on sentence verification | Mental rotation |

**TABLE 3.5**
**Deficits Reported in Studies of Turner Syndrome Subjects (continued)**

| | Study | N | %45X | Age(s) | Controls | Tests Given | Results | Deficits |
|---|---|---|---|---|---|---|---|---|
| 14. | Berch, Briscoe, Dignan, Kirtendall, & SMith (1985) | 10 | 160 | Mean=8.3 Yes | Rotation Task | | Comparable accuracy; Slower RTs; do not use rotation strategy | Mental Rotation |
| 15. | Pennington, Heaton, Karmark, Pendleton, Lehman, & Schucard (1985) | 10 | 100 | 15-33 | Yes | Halstead-Reitan Aphasia Screening Exam Story & Figure Memory | TS impaired in long-term Memory and visuo-spatial ability; no deficit on ability frontal tasks | Long-term memory; Visuo-spatial |
| 16. | McGlone (1985) | 11 | 73 | 13-18 | Yes | McGill Anomalies Money Closure Rey Ostereith Knox Cubes Wechsler Memory ITPA (Sequencing) Recurring Figures Spatial Relations L-R Discrimination Card Sorting Oral Fluency Continuous Performance Manual Task | TS poorer in Spatial Spatial Problems in absence of model; low scores on some tasks secondary to poor coordination | construction; Motor coordination |
| 17. | Lewandowski, Gostenbader, & Richman (1985) | 10 | 90 | 9-33 | Yes? | Ravens Benton Visual Retention Bruinincks | TS Problems in drawing and design copying; Clumsy, poor balance; dexterity and coordination problems | Visuo-motor integration Motor uncoordination |

TABLE 3.5
Deficits Reported in Studies of Turner Syndrome Subjects (continued)

| | Study | N | %45X | Age(s) | Controls | Tests Given | Results | Deficits |
|---|---|---|---|---|---|---|---|---|
| 18. | Nyborg & Nielsen (1981) | 34 | 41 | 15-33 | Yes | Rod & Frame Test<br>Embedded Figures<br>Money Road Map | TS significantly poorer on all tasks if untreated or treated > 2 years | Visual spatial;<br>Direction sense;<br>Numerical |
| 19. | Robinson et al. (1986) | 9 | 63 | 11-18 | Yes | Bruininck's<br>Spatial Relations<br>Visuo-motor Integration<br>Memory 3<br>Language Tests<br>Token Task | TS decreased perceptual awareness of body in space; Striking deficit in perceptual skills; poorer memory; Adequate receptive and expressive language but some difficulty following up verbal commands | Body awareness;<br>Perceptual Memory;<br>Sequencing |
| 20. | McCauley, Kay, Ito & Treder (1987) | 17 | 58 | 9-17 | Yes | Beery VMI<br>Embedded Figures<br>Affect Discrimination | TS significantly poorer on Beery and affective discrimination | Visuo-motor Integration<br>Facial affect Discrimination |

Pennington et al., 1985; Robinson et al., 1986; Silbert et al., 1977; Waber, 1979) were prone to report weaknesses also in the verbal, motor, and visuo-spatial areas. Because these verbal deficiencies appeared to be related more to the productive aspects of language processing (sequencing, fluency) than to language comprehension or quality of speech production, they might be related to the poorer organizational (Serra et al., 1978) and motor coordination skills (Lewandowski et al., 1985; McGlone, 1985; Robinson et al., 1986) also associated with the syndrome.

In summary, this analysis shows that while TS individuals demonstrate deficiencies across a broad spectrum of visuo-spatial skills, they also have difficulty in verbal sequencing, verbal fluency, motor coordination and organizational skills. Unfortunately this research is limited by either inadequate testing procedures, lack of controls, or small sample sizes. Furthermore, the "shotgun" approach to assessment using wide arrays of psychometric procedures signifies only a minor improvement over past work. Unless a more theoretically-guided approach is used that will survey the full complement of human cognitive processing, the TS deficit will continue to be as varied as the types of procedures selected by investigators.

## Cognitive Processing Mechanisms

The study of TS individuals has provided a rare opportunity to examine a relatively homogeneous group, which until now was believed to have a relatively pervasive, unitary cognitive deficit. It was thought that this information might be particularly useful for understanding the cognitive processing mechanisms underlying specific learning disabilities in the general population. Surprisingly, however, very little systematic research of this kind has been conducted on these individuals, possibly because of the difficulty in ascertaining TS samples and finding suitable controls.

In the late 1970s we undertook a study along these lines (Rovet & Netley, 1982). Our goal was to identify the components of cognitive processing that were contributing to the spatial deficit in TS females. Using laboratory-based chronometric procedures developed by cognitive psychologists to assess individual components of mental processing, we attempted to determine precisely how TS individuals processed verbal and spatial information. In one study, 31 TS subjects and 31 age- and VIQ-matched normal female controls were compared using a simplified version of the Shepard and Metzler (1971) mental rotation paradigm. This task required subjects to determine whether pairs of abstract block-like figures shown in different spatial orientations had the same or different three-dimensional configurations (see Figure 3.2). Data were analyzed as a function of the linear relationship between reaction times and the angular separations between pairs of stimuli. Intercept values were thought to reflect visual encoding processes, and slopes-mental rotation processes.

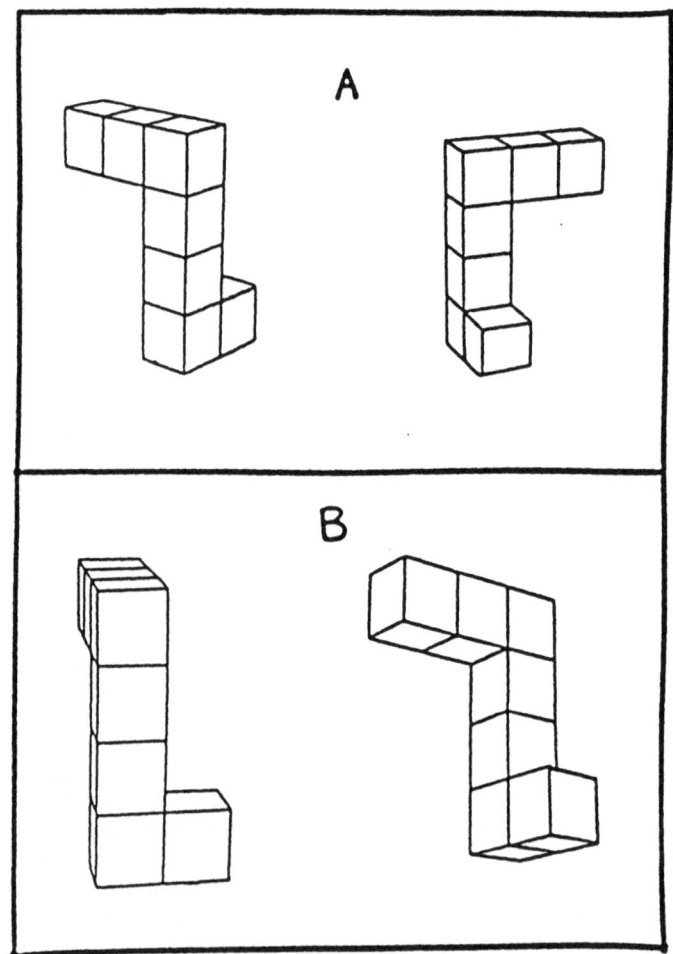

**Figure 3.2** Sample items from rotation task: A-"same" response; B-"different" response. (<u>Note</u>. From "Processing Deficits in Turner's Syndrome" by J. Rovet and C. Netley, 1982, <u>Developmental Psychology, 18</u>, p. 79. Copyright 1982 by the American Psychological Association, Reprinted by permission.)

Results showed that the TS subjects did significantly more poorly on this task and that their difficulties could be attributed mainly to the slope component of the reaction-time relationship. As can be seen in Figure 3.3, the TS subjects appeared to use the same mental rotation strategy as controls, because their performance also varied linearly with the angular distance separating stimuli. However, the TS subjects were far less efficient at carrying out the rotational process, as indicated by their steeper RT slope. They performed comparably to controls at visually encoding spatial information, as indicated by similar intercept values.

In a second experiment, 23 TS subjects and 23 matched controls were given a sentence verification task. They were asked to judge the veracity of a series of pictures showing a male and female cartoon figure engaged in activities such as kicking or chasing each other. Sentences printed below the picture varied in affirmative true, affirmative false, negative false, and negative true conditions. Analysis of the data in terms of the components of the linear relationships between sentence complexity and response times indicated that the groups did not differ in accuracy or processing strategy (Figure 3.4).

In a third experiment, the performance of a pair of twin sisters, one with TS, was contrasted on the same two tasks. The results showed the identical set of relationships found in the first two experiments. As can be seen in Figure 3.5, although the sisters performed identically on the sentence verification task, the non-TS twin outperformed the TS twin. Therefore, these results confirm that the TS deficit is mainly confined to the processing of spatial information, in particular the transformation of visual images, as opposed to the processing of verbal or propositional codes.

Berch and Kirkendall (1986) have also used a simplified version of the Shepard-Metzler task with TS girls averaging eight years of age. They found that while the girls did not differ from controls in accuracy, they failed to deploy the mental rotation strategy used by controls and did not take advantage of training to do so (presumably they used a verbal mediation strategy to solve the task). As in our study, Berch and Kirkendall found that the TS subjects responded much more slowly, which was thought to be reflective of their greater difficulties in visually encoding and comparing mental representations and/or in response execution. On a spatial visualization task requiring the children to judge whether pairs of puzzles (one assembled, one unassembled) matched, TS children showed no differences from controls in error rates or strategies. On a third task, spatial memory, which required the children to reconstruct the left-to-right spatial order of four pictures, Berch and Kirkendall found no difference between TS and control groups in accuracy or strategy. These authors suggest their results indicate a subtle dysfunction in spatial processing reflecting slower encoding proficiency and poorer transformational abilities. The ability to

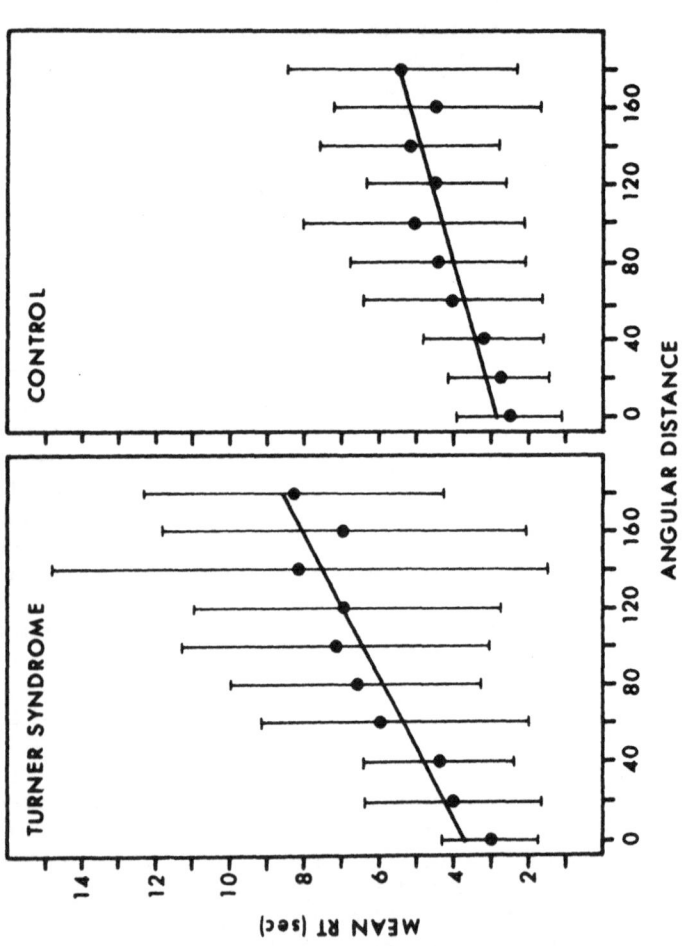

**Figure 3.3** Relationships between item angular distance and reaction time (RT). RT = 3.54 + .027x for Turner Syndrome subjects; RT = 2.88 + .015x for controls.) (Note. From "Processing Deficits in Turner's Syndrome" by J. Rovet and C. Netley, 1982, Developmental Psychology, 18, p. 82. Copyright 1982 by the American Psychological Association, Reprinted by permission.)

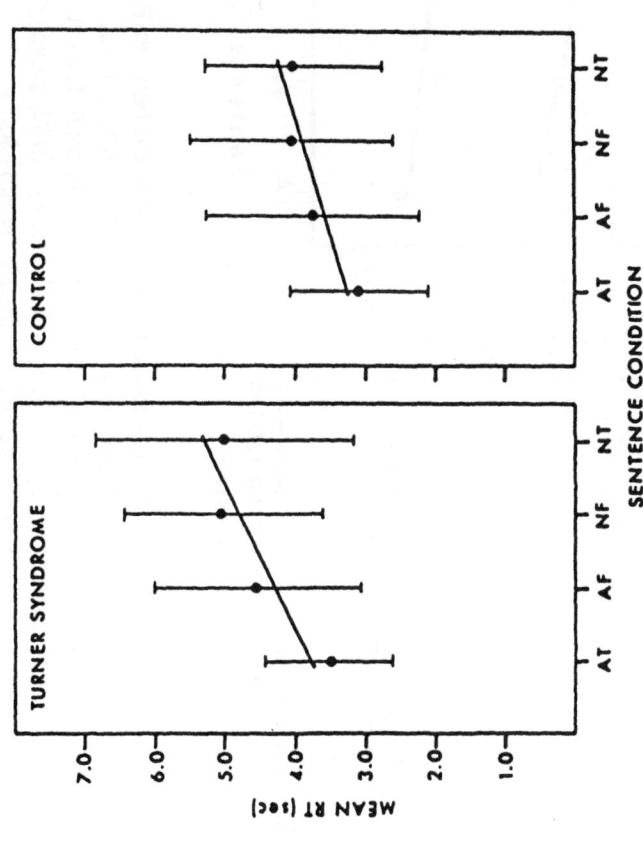

**Figure 3.4** Relationships between sentence condition and reaction time (RT). (RT = 3.28 + .499x for Turner syndrome subjects; RT = 2.93 + .309x for controls.) (Note. From "Processing Deficits in Turner's Syndrome" by J. Rovet and C. Netley, 1982, Developmental Psychology, 18, p. 87. Copyright 1982 by the American Psychological Association. Reprinted by permission.)

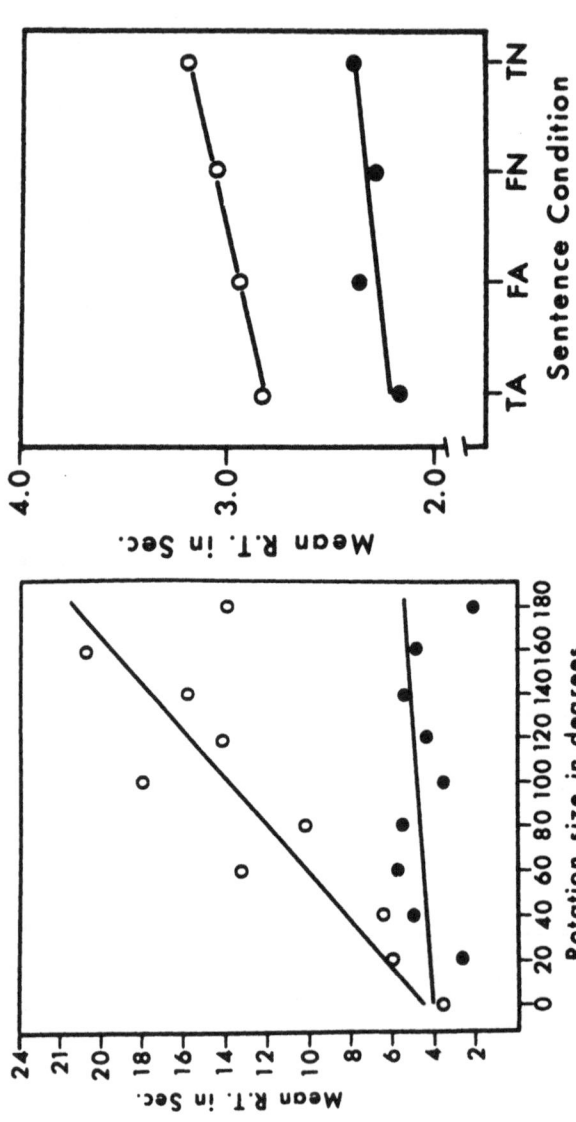

**Figure 3.5** Relationships between reaction time (R.T.) and rotation size (left panel) and between R.T. and sentence condition (right panel) for Turner syndrome (TS) subject (open circles) and her twin sister (filled circles). (Left panel: R.T. = 4.26 + .096x for TS twin; R.T. = 4.11 + .007x for non-TS twin. Right panel: R.T. = 2.69 + .12x for TS twin; R.T. = 2.17 + .05x for non-TS twin.) (Note. From "Processing Deficits in Turner's Syndrome" by J. Rovet and C. Netley, 1982, Developmental Psychology, 18, p. 89. Copyright 1982 by the American Psychological Association. Reprinted by permission.)

retain spatial information in short-term memory was comparable in the two groups.

In summary, the results of these more experimentally-based studies reflecting the abilities of actively encoding and transforming visuo-spatial information indicate that the TS deficit is highly selective. The ability to retain spatial information in memory or carry out mental transformations of verbally represented information appears to be unaffected. This type of research marks a first step in identifying the specific components of mental processing that may be contributing to the difficulties observed with psychometric testing. Further studies along these lines are clearly warranted.

## NEUROLOGICAL/NEUROPSYCHOLOGICAL CHARACTERISTICS

One of the major issues that continues to intrigue investigators regarding Turner syndrome concerns the impact of a missing X chromosome on brain function, both in fetal life and throughout development. Of particular interest are the effects of the lack of sex chromatin material on cellular differentiation of the brain, whether this occurs directly or indirectly through altered hormone levels or other biochemical factors, as well as the effects of altered hormone levels on subsequent brain function (see Nyborg, Chapter 5). (The converse question concerning how additional sex chromosome material affects brain development and functioning has been similarly explored in our laboratory and is dealt with by Netley in Chapter 6).

In reviewing the literature, the investigation of the neurological/neuropsychological concomitants of Turner syndrome appears to follows four lines of pursuit: (1) neurological studies, (2) neuropathological studies, (3) neuropsychological studies, and (4) hemispheric lateralization studies.

### Neurological Studies

The EEG studies carried out on TS women before 1974 have been reviewed by Nielsen, Nyborg, & Dahl (1977). The general consensus is that between 1/4 and 3/4 of the TS patients show atypical EEG wave patterns, but there is no consistency in wave form, severity, locus, or extent of localization. For example, Palm, Pfeiffer, Ammermann, and Schulte (1973) reported an overrepresentation of low amplitude and diffuse beta and alpha-beta mixture. They concluded there was a mild diffuse brain aberration in Turner syndrome. Poenaru, Stanesco, Poenaru, and Stoian (1970) studying 71 TS patients found only eight who had a normal EEG; they suggested that the EEG changes were due to immaturity of the cerebral cortex or disorders of the regulatory function of the mesencephalic reticular formation.

More recently, Tsuboi and Nielsen (1976) reported on the EEG findings of a group of TS patients. They found that 17 were normal, 17 had borderline results, eight had an unspecific abnormality, and one had a specific problem. Higher frequencies of paroxysmal EEG abnormalities were more evident among 45,X than other karyotype cases (39% versus 10%). Of the 17 cases with an increased amount of 14-18/s beta waves, 13 had diffuse and four had frontal dominant abnormalities. There were no indications of parietal lobe EEG aberrations or occipital lobe defects, which would be consistent with the cognitive data indicating a spatial impairment.

Although epilepsy was common among institutionalized TS patients, it was reported to be rare in TS individuals with normal intelligence (Nielsen et al., 1977). Hence, epilepsy does not appear to be a common feature of Turner syndrome.

### Neuropathological Studies

Two studies have reported on the results of three TS cases brought to autopsy. In the first, Brun and Skold (1968) described the findings for a 16-year-old TS girl who died due to repeated bleedings from a medulloblastoma. She was described as being mildly retarded and as having been previously hospitalized twice for epilepsy (petit mal variety). Repeated neurological examinations were generally unremarkable. Previous testing indicated IQ at the low end of normal, a deficiency on the Benton task of visual retention, but good school performance.

The autopsy results revealed a hemorrhagic tumor in the cerebellum (presumed cause of death), a right-sided pea sized gray nodule in the frontal horn of the ventricular system, a small grey heteroptopia in the white matter, and a normal gyral pattern on gross inspection. Microscopically, the outline of her cortex was unremarkable save for occasional cytoarchitectonic abnormalities. These reflected indistinct lamination from not all six laminae being recognizable and a varying width of the molecular layer. The latter was found to be hypercellular owing to too many glial cells in the outer layer and nerve cells scattered diffusely or gathered in nests. The white matter was also characterized as hypercellular with many glial cells and heterotropic nerve cells, as were mild cortical dysplasias and grey matter heteroptopias in the centrum semiovale. These findings appear to suggest a mild disturbance in neuronal migration, presumably occurring late in the histogenetical period of brain development.

Reske-Nielsen, Christensen, and Nielsen (1982) have described their findings on two TS patients who died at 52 and 24 years, respectively. The first case, a kindergarten teacher with normal intelligence, had a 45,X karyotype with typical Turner somatic features. She died from bronchopneumonia secondary to diabetes. Microscopic examination of the cerebrum showed atrophy and blurred leptomeninges presumed due to diabetes and epilepsy. There were no developmental brain abnormalities.

Case 2 with a 45X/46Xi (Xq) karyotype and typical Turner features had normal intelligence (VIQ = 104, PIQ = 90, FSIQ = 98) with a mild spatial impairment (Verbal Comprehension = 11, Perceptual Organization = 9). She was reported to have difficulties in arithmetic and the recall of names, was observed to be weak in visuo-motor integration and constructional organization, and slow in visual search and scanning. Partial left-sided neglect and difficulties on more complex visual tasks and verbal tasks involving spatial relationships were also evident. She died of a ruptured cerebral aneurysm.

The autopsy of Case 2 revealed brain atrophy with extremely small gyri in the junction of the temporo-parietal and occipital lobe regions. Microscopic examination revealed both acute and chronic changes, reflecting the effects of the rupture and structural lesions, particularly in the right middle cortex, secondary to an atherosclerotic process. However, there were no abnormalities in the laminar architecture, focal cortical dysplasias, or heteroptopias to suggest a developmental migration disturbance. Therefore, her poor spatial task performance appears to reflect decreased function due to aberrations in the posterior right hemisphere and not a developmental anomaly as in the Brun and Skold (1968) case.

These results signify three very different kinds of cerebral anomalies, each consistent with the individual's pattern of intellectual functioning. There is no consistency among the cases and no conclusive evidence to show that all TS women have an inborn localized developmental brain aberration (Reske-Nielsen et al., 1982). Clearly, further studies along these same lines are important, and concerted effort should be made for the collaboration of neuropsychologists and cytopathologists.

## Neuropsychological Studies

The research in this area falls into two major classifications: studies comparing TS individuals to patients having known localized brain lesions, and studies examining the performance of TS patients on clinical measures of brain dysfunction. One study (Pennington et al., 1985) has combined both approaches.

Comparisons with neurological patients. One of the earliest attempts at characterizing the neurological manifestations in the TS deficit was by Money (1973), who compared TS individuals to patients with right parietal lobe lesions or Gerstmann Syndrome. Of the 19 symptoms examined, TS females shared three with both of the other syndromes: visuo-constructional, right-parietal lobe disabilities, and numerical disabilities. However, they did not demonstrate visual neglect problems, exclusive to the right parietal syndrome group, nor aphasic and language difficulties, exclusive to the Gerstmann syndrome group. Money (1973) used this information to conclude that there may be a focal parietal defect in Turner syndrome which does not involve the language functions of the dominant hemisphere. He attributed this to a fetal neurohormonal mechanism affecting parietal lobe development.

Three recent neuropsychological investigations have failed to find support for Money's (1973) claim of a focal defect in the parietal lobes of TS individuals (McGlone, 1985; Silbert et al., 1977; Waber, 1979). Though based on smaller samples of TS patients, these studies have used fairly detailed and extensive test batteries. In Waber's study (n = 11), the cerebral dysfunction appeared to be generalized to both hemispheres and involved in parietal and frontal areas within the hemispheres. In contrast, Silbert et al. (1977) (n = 13) suggested a nonspecific deficit in cortical functioning lateralized totally to the right cerebral hemisphere, whereas McGlone (1985) (n = 11) suggested a greater localizing effect in the right post-central and adjacent parietal cortex.

The results of these three studies emphasize the considerable variability in the neuropsychological manifestations of TS. However, they do show a relationship between exclusivity of the spatial problem and localization of the neurocognitive impairment. Specifically, whenever such individuals demonstrated cognitive impairments that extended beyond the spatial domain, as in Waber's (1979) study, frontal and dual hemisphere involvement were more likely to be found.

Studies using clinical test batteries. There are three studies that have administered clinical neuropsychological test batteries to TS individuals. The first, by Kolb and Heaton (1975), involved only a single TS case who was given the Halstead-Reitan neuropsychological battery. Her Average Impairment Index, which fell in the impaired range, was suggestive of cerebral dysfunction strongly lateralized to the right hemisphere.

In the second study, Christensen and Nielsen (1981) conducted a neuropsychological investigation of 17 TS women using Luria's evaluation procedures. Their results indicated deficits on complex visual tasks involving sequencing and on motor coordination tasks, especially more complex motor patterns involving the left side of the body. These findings were interpreted as suggesting an impairment in the right posterior area, especially the temporo-parieto-occipital junction. An impairment corresponding to the basal areas was also noted.

The third study, by Pennington et al. (1985), involved a detailed assessment of 10 45,X right-handed TS subjects who were compared with 20 normal age-matched controls, 12 neurological patients with right hemisphere damage, 10 with left hemisphere damage, and 10 with diffuse lesions. All subjects received an expanded Halstead-Reitan battery, which included additional tests of frontal lobe functions, long-term memory, and motor coordination. The results revealed that the TS group was significantly more impaired than the normal controls and not unlike the three brain-damaged groups. In terms of lateralization of impairment, the TS subjects as a group were most similar to the diffuse-damage neurological patients and normal controls, although two of the ten TS subjects had a right lateralizing pattern of results. The neurocognitive deficit in the TS subjects was not found to be specific to any one psychological domain. Moderate impairments were also found in conceptual and attentional skill areas, nonverbal auditory processing, language fluency (reflecting calculation and right-left orientation) and memory. These results imply diffuse involvement of both hemispheres, with little if any involvement of the frontal lobes. The type of memory deficit noted in these individuals is suggestive of a dysfunction in the medial temporal and hippocampal structures.

In summary, five of the above studies indicate a definite right hemisphere dysfunction (Christensen & Nielsen, 1981; Kolb & Heaton, 1975; McGlone, 1985; Money, 1973; Silbert et al., 1977) while two indicate bilateral effects (Pennington et al., 1985; Waber, 1979). It is important to continue this line of investigation to determine whether there is indeed bilateral involvement which was missed in the right-hemisphere-only studies due to Type II statistical errors, or whether Waber's and Pennington's results were due to the unique and peculiar characteristics of the particular subjects they sampled.

The seven studies were even more varied as to localization effects. Only two (Kolb & Heaton, 1975; Silbert et al., 1977) reported similar phenomena, a generalized right hemisphere deficit. The types of specific deficits ranged from right parietal (Money, 1973), right posterior (Christensen & Nielsen, 1981), and right post-central and parietal (McGlone, 1985) to right and left parietal and frontal (Waber, 1977) and right and left medial temporal and hippocampal (Pennington et al., 1985). One conclusion that can be drawn from these studies is that regardless of additional involvement, right parietal development is consistently affected in TS individuals.

## Hemispheric Lateralization Studies

A number of studies have tried to determine whether the characteristic cognitive impairment of TS might reflect atypical hemispheric specialization and organization. Table 3.6 summarizes the results of six published studies that have used a variety of different laterality techniques. Five studies used dichotic listening procedures with verbal stimuli, while one (McGlone, 1985) used a tachistoscopic visual half-field presentation of verbal (letters) and nonverbal (dot enumeration) stimuli. Thus all of the tasks were assessing for left hemisphere or right ear/right visual field superiority, except McGlone's (1985).

The results indicate that TS subjects consistently show either weaker left hemisphere asymmetries or reverse asymmetries (i.e., right hemisphere biases for verbal information). In three studies, degree of appropriate lateral bias was associated with better spatial ability. The one task assessing right hemisphere lateralization for nonverbal

**TABLE 3.6**
**Results of Hemispheric Lateralization Studies**

| | Study | N | Controls | Tasks | Results | Association With Ability |
|---|---|---|---|---|---|---|
| 1. | Netley (1977) | 14 | Yes | Dichotic Digits | LEA in 57% TS vs. 16% C | LH superiority correlated with less spatial impairment (r = .721, p < .05) |
| 2. | Waber (1979) | 10 | Yes | Dichotic Phonemes | LEA in 40% TS vs. 0% C | LEA not associated with spatial ability |
| 3. | Gordon & Galatizer (1980) | 14 | Yes | Dichotic Digits Dichotic Words | No difference on digits TS—strong REA for words | |
| 4. | Netley & Rovet (1982) | 35 | Yes | Dichotic Digits Triads | LEA in 17% TS vs. 0% C REA in 29% TS vs. 43% C | Lack of ear advantage associated with spatial impairment in TS only |
| 5. | Lewandowski et al. (1985) | 9 | Yes | Dichotic Syllables Verbal Concurrent Matching Geometric Concurrent Matching | LEA in 55% TS on syllables | |
| 6. | McGlone (1985) | 10 | Yes | T-Scope Letters T-Scope Dot Enumeration | RVF superiority for letter recognition in controls only; TS < C in RVF, not LVF No effect for dots | |

stimuli (McGlone, 1985) did not show a difference between TS and control groups.

We have recently had the opportunity to examine hemispheric lateralization in another set of 15 TS subjects who were given four lateral asymmetry tasks assessing both left and right hemispheric specialization. The tasks presumed to measure left hemisphere specialization involved the dichotic presentation of stop consonants and a tachistoscopically presented letter recognition procedure; the two presumed to assess right hemisphere specialization included a dichotic listening task with musical stimuli and a T-scope dot enumeration task. Controls consisted of a standardization sample of 118 normal females.

As shown in Figure 3.6, TS subjects demonstrated weaker or reverse asymmetries than controls for the two left hemisphere tasks but indicated greater left hemisphere involvement for the two tasks presumed to measure right hemisphere specialization. The mean individual right and left side scores are depicted in Figure 3.7 for each of the four tasks. It can be seen that for the two verbal, or presumed left hemisphere tasks (panels a and b), TS subjects obtained either equivalent or lower right-side than left-side scores. In contrast, controls scored consistently better on the right side. On the two presumed right hemisphere tasks, where a left-side advantage would be expected (see panels c and d), TS subjects showed the reverse effect, obtaining higher right- than left-side scores. This was more dramatic for the task involving musical stimuli than the task requiring dot enumeration. Together these findings appear to suggest that TS individuals may be using their left hemisphere to a greater degree than normal in processing nonverbal information, thereby compromising its potential in exclusively processing verbal information.

Finally, the results are reported for a task of asymmetrical cognitive performance developed by Gordon and colleagues (Harness, Epstein, & Gordon, 1984) known as the Cognitive Laterality Battery. This test battery contrasts one's performance on presumed right hemisphere cognitive tasks (e.g., orientation of 2- and 3-dimensional objects, localization of points in space, perception of incomplete figures) with one's performance on presumed left hemisphere cognitive tasks (serial processing of sounds, numbers and verbal instructions, verbal fluency), as compared with a large standardization sample. Reporting on 11 TS patients, Galatzer and Gordon (1982) noted a bias in favor of left hemisphere functions among these individuals.

In summary, these studies indicate that when TS individuals are assessed with tasks of lateralized hemispheric specialization, they show weaker left hemisphere advantages for verbal information and reverse (i.e., left hemisphere) advantages for nonverbal information. Degree of lateralization appears to vary with type of stimulus and mode of presentation.

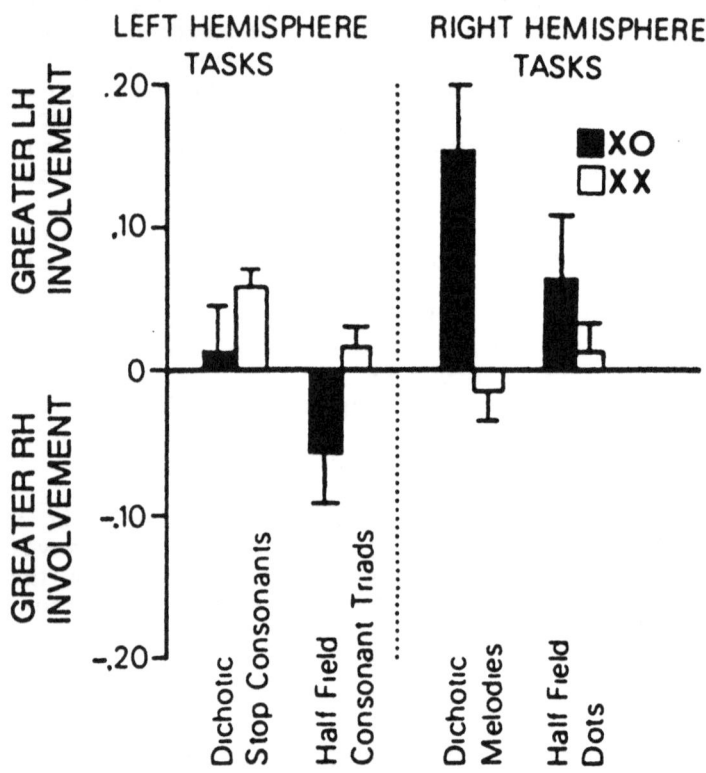

**Figure 3.6** Mean phi scores (degree of left hemisphere [LF] or right hemisphere [RH] involvement) on laterality tasks for Turner (XO) and control (XX) subjects.

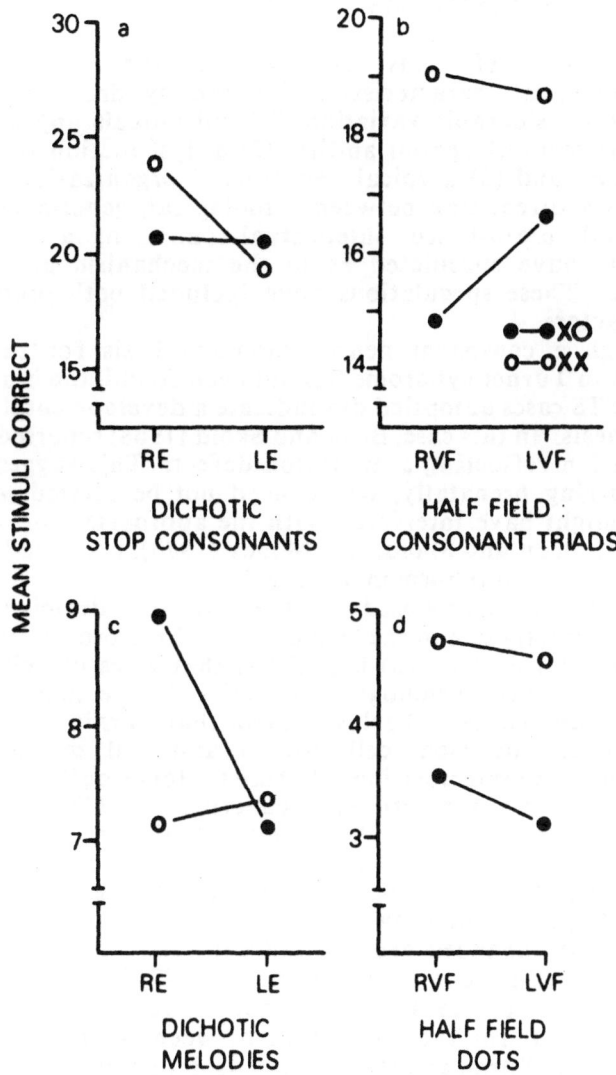

**Figure 3.7** Performance in relation to side of presentation (right ear [RE] or left ear [LE]; right visual field [RVF] or left visual field [LVF]) on four laterality tasks for Turner (XO) and control (XX) subjects.

## BIOLOGICAL MECHANISMS

The results of this review on the cognitive and neuropsychological characteristics of Turner syndrome have shown that despite considerable variation, TS individuals appear to have (1) an impairment of spatial ability, (2) a dysfunction in the right parietal lobes, and (3) atypical hemispheric organization. Because this implies a direct link between a biological, genetic sex-related disorder and brain-based intellectual function, a number of investigators have speculated as to the mechanism affecting TS individuals. These speculations have included both prenatal and postnatal factors.

Although a consistent neuroanatomical basis for the spatial impairment in Turner syndrome has not been found, the brain of one of the three TS cases autopsied did indicate a developmental anomaly of neurogenesis. In this case, Brun and Skold (1968) reported a defect in laminization reflecting a migration defect. This suggests that a factor occurring prenatally, which need not be related to Turner syndrome, might have interfered with the appropriate development of the brain. Brun and Skold, like Money (1973), presumed that this was under direct neurohormonal control.

Netley (1977) has proposed that the pattern of deficit in Turner syndrome is directly due to the lack of an X chromosome. His theory follows from Barlow's (1973) hypothesis that rates of cell division underlie the neuropsychological deficits in persons with sex chromosome anomalies. Barlow noted that "Brain development depends on cell division, cell growth and cell migration, and disturbances in the interrelations of these factors would undoubtedly lead to disturbed brain structure and function" (p. 112). Polani (1977) also promoted this theory, while Netley (1977) has subsequently expanded it to account for a wider range of findings. In particular, Netley has proposed that the reduced chromosome complement in Turner syndrome would allow for faster cell division rates, thereby contributing to enhanced growth. In contrast, the additional sex chromosome complement in children with 47,XXY and 47,XXX karyotypes would lead to slower than normal cell division rates and, therefore, account for the association between the delayed growth rate of extra X children and their verbal deficits (see Netley, Chapter 6). Thus according to Netley, it is the increased rate of cell division in Turner syndrome that contributes to the atypical development of the right hemisphere during the neonatal period. This would be reflected in a generalized or diffuse pattern of right hemisphere dysfunction, as noted in a number of studies.

In addition to the effect of Turner syndrome on early brain development, it is plausible that other factors expressed in later development are contributing to the impairments of TS individuals, such as lack of endogenous sex steroids (Waber, 1979), exogenous hormone administration (Nyborg & Nielsen, 1977), or factors associated with high gonadotropin levels (Galatzer & Gordon, 1982).

Differences in one or more of these factors may account for the large degree of variability observed among the patients studied. It is not clear whether they produce structural lesions (Reske-Nielsen et al., 1982, Case 2) or result in cerebral dysfunctions due to improper neurohumoral/neurotransmitter activation. However, there has been very little research conducted in this area, and certainly more studies are needed concerning the role of sex hormones and their neuroregulatory functions in cognitive processing (e.g., Kimura, 1987).

Finally, a recent single case study by myself (Rovet, 1987) suggests the possibility of an alternative, somewhat more dynamic explanation involving interhemispheric interactions. In this study, I compared the performance of a TS patient before and after developing left temporal lobe seizures. Her results are shown in Table 3.7. Prior to epilepsy, she showed the typical TS ability profile with a pronounced spatial disadvantage relative to verbal skills. When assessed shortly after her first tonic clonic seizure involving a left hemisphere focus, she demonstrated a marked improvement in spatial ability, an effect which is not typically observed with chromosomally normal individuals. This included perfect performance on the Coding subtest in less than the allotted time. Her verbal skills were unaffected, despite the left-sided focus. As shown in Table 3.8, laterality testing revealed decreased left hemisphere involvement in verbal processing and increased right hemisphere involvement in nonverbal processing following epilepsy. In other words, her neurocognitive functions appeared to have normalized after she sustained left hemisphere damage from epilepsy.

**TABLE 3.7**
**Intelligence Test Scores of Turner Syndrome Patient Before and After Seizure Disorder**

|  | Before | After |
|---|---|---|
| **WECHSLER** | | |
| Verbal IQ | 106 | 104 |
| Performance IQ | 81 | 112 |
| **PMA** | | |
| Verbal IQ | -- | 100 |
| Performance IQ | -- | 111 |

Although a number of explanations for this highly unusual finding are possible, including test-retest effects and the impact of medication, they cannot adequately account for an improvement of this magnitude in spatial ability. A recent argument by Kinsbourne (1982) seems to fit these results better. According to Kinsbourne, the original spatial deficit may have been due to an inhibitory influence of the left hemisphere on the right, which would interfere with her ability to process spatial information. Presumably this would have been relieved by the epilepsy and the subsequent left hemisphere damage, thereby freeing her to use her right hemisphere to its full potential. Although these "inhibitory factors" as well as their biochemical bases are unknown, it is not unreasonable to presume that this may be a useful direction for future research on the biochemical bases of brain function.

**TABLE 3.8**
**Laterality Test Results Before and After Seizure Disorder**

|                  | *phi*-score |        |
|------------------|-------------|--------|
|                  | Before      | After  |
| Laterality Task  |             |        |
| A. Left          | .067        | -.039  |
| B. Right         | .140        | .193   |

Note.   + *phi* score reflects an LH advantage for line A
+ *phi*-score reflects an RH advantage for line B

## CONCLUSIONS

The results of this review have shown that individuals with Turner syndrome do have a cognitive deficit, but it appears to be more heterogeneous than originally believed. Although there is some evidence to indicate that it is neurologically mediated, there is no general consensus as to its etiology or its neurological basis. Inconsistencies and inconclusive findings appear to abound in the literature. In integrating the published research on this topic, it was evident that the majority of studies were either poorly founded or methodologically unsound; if better controlled as in the more recent work, sample sizes were very small, increasing the likelihood of missing effects that really existed. Furthermore, there also appeared to be a serious need for a more theoretically-oriented approach to the study of specific cognitive deficits in order to identify better the source or sources of the impairment.

Clearly, more research on this interesting and unusual population of individuals is required. However, because of the difficulty in ascertaining sufficient sample sizes to conduct well-controlled, methodologically-sound studies on these patients, it is recommended that a multicenter collaborative approach be considered for future investigation in order to optimally address the outstanding questions. A first step in coordinating data consists of collaborative attempts exemplified by the symposium on which this book is based, and conferences such as the March of Dimes sponsored meetings for reporting on the development of neonatally-identified children with sex chromosome anomalies. However, the next phase of the work must be marked by fundraising and development of procedures so that teams at different hospitals can ascertain the same data and use comparable controls. One possibility is linking this research with multicenter growth hormone studies that are ongoing in the United States and Canada.

Collaborative research requires attracting the interest of more investigators, which it is hoped this chapter will achieve. When the important outstanding research questions are addressed in this way, only then can the information be obtained that is needed by parents or prospective parents of a TS child.

## NOTE

1. Studies 1, 3, and 6 are not included in this analysis because their participants are presumably also described in the larger studies by the same authors (i.e., 2 and 12). Studies 7, 9, and 11 are not included because they do not present the actual IQ scores.

## ACKNOWLEDGMENTS

Support from two Canadian agencies, the Ontario Mental Health Foundation and Health and Welfare Canada (a National Health Research Development Program Scholarship) have made possible the research described in this paper and its writing. I would like to thank John Bailey and Jack Holland for their cooperation and continued referrals, Chuck Netley for his collaboration in the various studies, Donna Sorbara for psychometric testing, Jeannine Pinnsonneault for data analysis, laterality testing and investigation of controls, Betty Adamo for her assistance, and JoAnne Finnegan for her thorough and thoughtful critique of the manuscript. As well, I am very appreciative of my close friendship with Susan Charney and my many involvements in the Turner Syndrome Society that have evolved from the research. Both have made me aware of the "real" variability in the syndrome, the underlying talents that are never assessed, and the remarkable adaptability of these individuals in their daily living.

## REFERENCES

Alexander, D., & Money, J. (1966). Turner's syndrome and Gerstmann's syndrome: Neuropsychologic comparisons. Neuropsychologia, 4, 165-273.

Alexander, D., Walker, H. T., & Money, J. (1964). Studies in direction sense. Archives of General Psychiatry, 10, 337-339.

Banantyne, A. (1974). Diagnosis: A note on recategorization of the WISC scaled scores. Journal of Learning Disabilities, 7, 272-274.

Barlow, D. (1973). The influence of inactive chromosomes on human development. Humangenetik, 17, 105-136.

Berch, D. B., Kirkendall, K. L., Briscoe, G., Dignan, P. S. J., & Smith K. L. (1985). Spatial information processing in children with Turner's syndrome. Presented at the Society for Research in Child Development meeting, Toronto.

Berch, D. B., & Kirkendall, K. L. (1986). Spatial information processing in 45 X children. Presented at the American Association for the Advancement of Science meeting, Philadelphia.

Brun, A., & Skold, G. (1968). CNS malformations in Turner's syndrome: An integral part of the syndrome? Acta Neuropathologica, 10, 159-161.

Buckley, F. (1971). Preliminary report on intelligence quotient scores of patients with Turner's syndrome: A replication study. British Journal of Psychiatry, 119, 513-514.

Christensen, A. L., & Nielsen, J. (1981). Neuropsychological investigation in 17 women with Turner's syndrome. In W. Schmid and J. Nielsen (eds.) Human behaviour and genetics, Elsevier/North Holland Biomedical Press, Amsterdam.

Cohen, H. (1962). Psychological test findings in adolescents having ovarian dysgenesis. Psychosomatic Medicine, 24, 249-256.

Cohen, J. (1957). A factor analytically based rationale for the Wechsler Adult Intelligence Scale. Journal of Consulting Psychology, 21, 451-457.

Dellantonio, A., Lis, A., Saviolo, N., Rigon, F., & Tenconi, R. (1981). Spatial performance and hemispheric specialization in Turner's syndrome. Clinical Genetics, 20, 370-371.

Galatzer, A., & Gordon, H. W. (1982). Cognitive asymmetries in patients with hormone and genetic abnormalities. Presented at the European Conference of The International Neuropsychological Society, Deauville, France.

Garron, D. C. (1969). Sex-linked recessive inheritance of spatial and numerical abilities and Turner's syndrome. Psychological Review, 77, 147-152.

Garron, D. C. (1977). Intelligence among persons with Turner's syndrome. Behavior Genetics, 7, 105-127.

Garron, D. C. (1978). Comment on "Spatial and temporal processing in patients with Turner's syndrome. Behavior Genetics, 8, 289-290.

Garron, D. C., Molander, L., Cronholm, B., & Lindsten, J. (1973). An explanation of the apparently increased incidence of moderate mental retardation in Turner's syndrome. Behavior Genetics, 3, 37-43.

Garron, D. C., & Vander Stoep, L. R. (1969). Personality and intelligence in Turner's syndrome. Archives of General Psychiatry, 21, 339-346.

Gordon, H. W., & Galatzer, A. (1980). Cerebral organization in patients with gonadal dysgenesis. Psychoneuroendocrinology, 5, 235-244.

Grumbach, M. M., VanWyck, J. J., & Wilkins, L. (1955). Chromosomal sex in gonadal dysgenesis relationship to male pseudohermaphroditism and theories of human sex differentiation. Journal of Clinical Endocrinology, 15, 1161-1193.

Harness, B. Z., Epstein, R., & Gordon, H. W. (1984). Cognitive profile of children referred to a clinic for reading disabilities. Journal of Learning Disabilities, 17, 346-351.

Kaufman, A. S. (1979). Intelligent testing with the WISC-R. New York: Wiley.

Kimura, D. (1987). Biological contributions to cognitive formation: The role of brain organization, sex and hormones. Neurocognitive Seminars, Baycrest Hospital, Toronto.

Kinsbourne, M. (1982). Hemispheric specialization and the growth of human understanding. American Psychologist, 37, 411-420.

Kolb, J. E., & Heaton, R. K. (1975). Lateralized neurological deficits and psychopathology in a Turner syndrome patient. Archives of General Psychiatry, 32, 1198-1200.

Lewandowski, L., Costenbader, V., & Richman, R. (1985). Neuropsychological aspects of Turner's syndrome. International Journal of Clinical Neuropsychology, 7, 144-149.

McCauley, E., Kay, T., Ito, J., & Treder, R. (1987). The Turner's syndrome: Cognitive deficits, affective discrimination and behavior problems. Child Development, 58, 464-473.

McGlone, J. (1985). Can spatial deficits in Turner's syndrome be explained by focal CNS dysfunction or atypical speech lateralization. Journal of Clinical and Experimental Neuropsychology, 7, 375-394.

Money, J. (1963). Cytogenetic and psychosexual incongruities with a note on space-form blindness. American Journal of Psychiatry, 119, 820-827.

Money, J. (1964). Two cytogenetic syndromes: Psychologic comparisons: I. Intelligence and specific-factor quotients. Journal of Psychiatric Research, 2, 223-231.

Money, J. (1973). Turner's syndrome and parietal lobe functions. Cortex, 9, 385-393.

Money, J., & Alexander, D. (1966). Turner's syndrome: Further demonstration of the presence of specific cognitional deficiencies. Journal of Medical Genetics, 3, 47-48.

Money, J., & Granoff, D. (1965). IQ and the somatic stigmata of Turner's syndrome. American Journal of Mental Deficiency, 70, 69-77.

Netley, C. (1977). Dichotic listening of callosal agensis and Turner's syndrome. Language Development and Neurological Theory, 11, 133-143.

Netley, C., & Rovet, J. (1982). Atypical hemispheric lateralization in Turner's syndrome subjects. Cortex, 18, 377-384.

Netley, C., & Rovet, J. (1983). Relationships among brain organization, maturation rate and the development of verbal and nonverbal ability. In S. J. Segalowitz (Ed.), Language functions and brain organization, New York: Academic Press.

Nielsen, J., Nyborg, H., & Dahl, G. (1977). Turner's syndrome. Acta Jutlandica XLV, 1, 190.

Nyborg, H., & Nielsen, J. (1977). Sex chromosome abnormalities and cognitive performance: III. Field dependence, frame dependence and failing development of perceptual stability in girls with Turner's syndrome. The Journal of Psychology, 96, 205-211.

Nyborg, H. (1986). Sex chromosomes, sex hormones and developmental disturbances: In search of a model. Presented at The American Association for the Advancement of Science meeting, Philadelphia.

Nyborg, H., & Nielsen, J. (1981). Sex hormone treatment and spatial ability in women with Turner's syndrome. In W. Schmid and J. Nielsen (Eds.). Human Behavior and Genetics, (pp. 167-182). Amsterdam: Elsevier Press.

Palm, D., Pfeiffer, R. A., Ammermann, M., & Schulte, H. (1973). EEG-Befunde bei Turner-Syndrome. Mschr. Kinderheilk, 121, 289-292.

Pennington, B., F., Bender, B., Puck, M., Salbenblatt, J., & Robinson, A. (1982). Learning disabilities in children with sex chromosome anomalies. Child Development, 53, 1182-1192.

Pennington, B. F., Heaton, R. K., Karzmark, P., Pendleton, M. G., Lehman, R., & Schucard, D. W. (1985). The neuropsychological phenotype in Turner syndrome. Cortex, 21, 391-404.

Poenaru, S., Stanesco, V., Poenaru, L., & Stoian, D. (1970). EEG dans le syndrome de Turner. Acta Neurologica Belgica, 70, 509-522.

Polani, P. E. (1961). Turner's Syndrome and allied conditions. British Medical Bulletin, 17, 200-205.

Polani, P. E. (1977). Abnormal sex chromosomes, behaviour and mental disorder. In Tanner J. (Ed.). Developments in psychiatric research, London: Hoddler & Stughton, pp. 89-128.

Reske-Nielsen, E., Christensen, A. L., & Nielsen, J. (1982). A neuropathological and neuropsychological study of Turner's syndrome. Cortex, 18, 181-190.

Robinson, A., Bender, B. G., Borelli, J. B., Puck, M. H., Salbenblatt, J. A., & Winter, J. S. D. (1986). Sex chromosomal aneuploidy: Prospective and longitudinal studies. Birth Defects: Original Articles Series, 22, 23-71.

Rovet, J., & Netley, C., (1982). Processing deficits in Turner's syndrome. Developmental Psychology, 18, 77-94.

Rovet, J. F. (1987). Improved spatial ability correlated with left hemisphere dysfunction in a patient with Turner's syndrome: Implications for mechanism. Unpublished manuscript.

Serra, A., Pizzamiglio, L., Boari, A., & Spera, S. (1978). A comparative study of cognitive traits in human sex chromosome aneuploids and sterile and fertile euploids. Behavior Genetics, 8, 143-154.

Shaffer, J. W. (1962). A specific cognitive deficit observed in gonadal aplasia (Turner's syndrome). Journal of Clinical Psychology, 18, 403-406.

Shepard, R. N., & Metzler, J. (1971). Mental rotation of three-dimensional objects. Science, 171, 701-702.

Silbert, A., Wolff, P. H., & Lilienthal, J. (1977). Spatial and temporal processing in patients with Turner's syndrome. Behavior Genetics, 7, 11-21.

Simpson, J. L. (1975). Gonadal dysgenesis and abnormalities of the human sex chromosomes: Current status of phenotypic-karotypic correlations. Birth Defects: Original Article Series, 11, 23-55.

Theilgaard, A. (1972). Cognitive style and gender role in persons with sex chromosome aberrations. Danish Medical Bulletin, 19, 176-286.

Theilgaard, A., & Philip, J. (1975). Concurrence of Turner's syndrome and anorexia nervosa. Acta Psychiatric Scandinavica, 52 31-35.

Tsuboi, R., & Nielsen, J. (1976). Electroencephalographic examination of 50 women with Turner's syndrome. Acta Neurologica Scandinavica, 54, 359-365.

Waber, D. P. (1976). Neuropsychological analysis of spatial ability in Turner's syndrome: Genetic implications. Paper presented at The American Psychological Association meeting, Washington, D. C.

Waber, D. P. (1979). Neuropsychological aspects of Turner's syndrome. Developmental Medicine and Child Neurology, 21, 58-70.

# 4 Psychosocial and Emotional Aspects of Turner Syndrome

Individuals with Turner syndrome (TS) provide a unique opportunity to study the impact of sex chromosome anomalies on subsequent growth and development. This chapter will review research on the psychological development of women with Turner syndrome. The focus will be on those papers that have attempted to characterize and assess emotional and psychosocial functioning. The goals of the chapter are to provide a comprehensive overview of the literature, assess the consistency of findings across studies and samples, and explore explanatory models. Much of the behavioral research conducted with TS women has focused on cognitive abilities. Cognitive and neuropsychological functioning are covered in the preceding chapter by Rovet and will not be reviewed here.

There are at least twenty-one investigations of the social and emotional development of women with Turner syndrome as well as many individual case reports. Although the individual case reports are of great clinical and heuristic value, this chapter will concentrate on those investigations that have evaluated groups of women with Turner syndrome in an effort to identify characteristic behavioral patterns.

Before evaluating specific studies, I will review some of the methodological problems that complicate any study in this area. First, are the issues related to sampling. Since sample recruitment is difficult, most studies consist of small samples of subjects representing a wide age range. Selection bias is also an issue since TS women who are known to medical centers and are willing to participate in research may not be representative of the overall population of women with this syndrome. Furthermore, we now know that a variety of discrete chromosomal anomalies are included under the umbrella of Turner syndrome. Although the 45,X karyotype remains the most common, most TS samples are comprised of women with a variety of karyotypes including mosaic cell lines. Samples of women with Turner syndrome are therefore more heterogeneous than is ideal, and small sample size prohibits detailed analyses of more homogeneous subgroups.

Second, the large number of physical anomalies associated with this syndrome can set an affected individual apart in terms of physical presentation and/or physical well-being. Short stature and gonadal dysgenesis are the most universal features, but some individuals are also affected by physical anomalies such as webbing of the neck, shield chest, cardiac and/or kidney defects. Thus, in contrast to some of the other sex chromosomal anomalies, individuals with Turner syndrome are usually identifiable by their physical presentation, thereby precluding blind evaluation within a research study. Physical appearance and health also affect personality and social development and thus confound assessment in these areas. With these precautions in mind let us move on to review the behavioral data available on individuals with Turner syndrome.

## GENDER IDENTITY AND SEXUAL BEHAVIOR

Psychosocial research efforts have focused primarily on the topics of gender identity formation, psychopathology, and personality characteristics. Interest in the sexual differentiation and gender identity of TS women was initiated by researchers investigating the contribution of various biological factors to psychological gender differentiation. Furthermore, before karyotyping techniques were perfected it was mistakenly thought that these women were genetic males. Consequently, TS women provided an unusual opportunity to study how incomplete or atypical formation of the sex chromosome complement and subsequent failure of gonadal differentiation affect sex role development, gender identity, and sexual functioning.

Ten studies were identified that have examined these issues, typically along with other psychosocial questions. However, assessment of sexual behavior was somewhat cursory in many of these studies, and control groups were included in only three. In spite of these methodological limitations across studies, as shown in Table 4.1, clear and unambiguous female gender identification has been reported consistently (Ehrhardt, Greenberg, & Money, 1970; Hampson, Hampson, & Money, 1955; Money & Mittenthal, 1970; Nielsen & Sillesen, 1981; Nielsen, Nyborg, & Dahl, 1977; Sabbath, Morris, Menzer-Benaron, & Sturgis, 1961; Shaffer, 1963; Taipale, 1979; Theilgaard, 1972). Girls with Turner syndrome are described as following a very traditional pattern of female development with a strong interest in child care. Ehrhardt et al. (1970) conducted the most extensive evaluation of gender identification and gender role. The 15 girls with Turner syndrome in this study, ages 8 to 14 years, exhibited patterns of conventional or stereotyped female behavior more frequently and had a significantly lower incidence of tomboyism during childhood than a control group matched for age and socioeconomic level.

While findings have also consistently supported a heterosexual orientation for developing girls and women with Turner syndrome, sexual drive in adult women has been described as low (Garron & Vander Stoep, 1969; Hampson et al., 1955). However, the data on sexual functioning are both less adequate and less consistent across studies than those on gender-related issues. Most samples include

**TABLE 4.1**
**Studies Including Some Assessment of Gender Identity and/or Sexual Behavior in TS Women**

| | Study | Sample Size | Age Ranges | Country | Controls | Measures* | Results |
|---|---|---|---|---|---|---|---|
| 1. | Hampson, Hampson, & Money (1955) | 11 | 9-27 | U.S. | None | Interviews Thematic Apperception Test (TAT) | Female GI in all cases; Hetero-fantasies; 5/5 older Ss reported evidence of sexual behavior |
| 2. | Sabbath et al. (1961) | 7 | 14-20 | U.S. | None | Interviews | Denial of sexual concerns in most cases; 2/7 expressed concern regarding sexual function |
| 3. | Cohen (1962) | 10 | 14-21 | U.S. | N=9 girls 13-17 yrs of age for DAP only | Draw-A-Person (DAP), TAT, Rorschach | Immature DAP and TAT responses in relationship to sexual feelings and development |
| 4. | Shaffer (1963) | 13 | 13-23 | U.S. | Questionnaire norms test | Guilford-Zimmerman Temperament Survey(GZTS) | Sign. Lower (more feminine) than normative group on GZTS Masculinity-Femininity Scale |
| 5. | Ehrhardt, Greenberg, & Money (1970) | 15 | 5-16 | U.S. | N=15 matched f/age, sex, race, IQ, and SES | Semistructured interview, DAP Lynn Structured Doll Play | TS > Controls - Interest in feminine appearance TS < Controls - Athletic interests, aggression |
| 6. | Money & Mittenthal (1970) | 73 | 10-Adult | U.S. | | Chart Review Interview | Female GI & GR in all cases; no homosexual fantasies or behaviors |

**TABLE 4.1**
**Studies Including Some Assessment of Gender Identity and/or Sexual Behavior in TS Women (continued)**

| | Study | Sample Size | Age Ranges | Country | Controls | Measures* | Results |
|---|---|---|---|---|---|---|---|
| 7. | Theilgard (1972) | -- | -- | Denmark | Women with one amenorrhea | TAT, DAP Word/ Association | Female GI & GR Immature DAPs |
| 8. | Nielsen, Nyborg, & Dahl (1977) | 45 | 7-39 | Denmark | 15 short stature amenorrheic, normal karyotype women (SS); 46 sisters of TS ± 5 years | Interviews | Female GI & GR in all cases; less frequent/satisfactory sexual relations for TS and SS controls than sister group |
| 9. | Taipale (1979) | 49 | 9-22 | Finland | None | Interviews DAP Draw-A-Tree Rorschach | Female GI in all; GR female with one tomboy |
| 10. | McCauley, Sybert, & Ehrhardt (1986) | 30 | -- | U.S. | None | Semistructured interviews | Heterosexual orientation but limited dating and/or sexual experience |

*Only measures relevant to this table are listed.

children, teenagers, and young adults and report only the prevalence of sexual fantasies and activities such as masturbation and intercourse in their samples. The data suggest that heterosexual romantic fantasies are common among TS females (Hampson et al., 1955; Money & Mittenthal, 1970), and some of these women also report erotic sexual fantasies and sexual behaviors (Hampson et al., 1955; Pollock, 1955; Sabbath et al., 1961). Sexual fantasies and behaviors (masturbation) are infrequently reported before the adolescent years and appear to increase with the onset of estrogen replacement therapy (Garron & Vander Stoep, 1969).

Nielsen et al. (1977) compared the social and sexual functioning of 45 TS females (28 eighteen years or older) to that of two control groups: short-statured, amenorrheic women with no chromosomal anomaly, and sisters of the TS women. They found less frequent and less satisfactory sexual relationships in their TS and short-statured samples than in the control groups of unaffected sisters. Fifteen of the 28 TS women had never had sexual intercourse; of the 13 others, ten felt their sexual relationships were satisfactory, and three had difficulties in this area. Similar results were obtained in our own study of adult TS women (McCauley, Sybert, & Ehrhardt, 1986). Twenty-seven of the 30 women sampled completed a detailed interview covering social and emotional development. These women reported only limited heterosexual contacts, a tendency to begin dating later than their peers, a later age of first sexual experience than peers, and a low frequency of sexual activity. Three of the nine women who were or had been married described significant sexual dysfunction, with problems ranging from low libido to symptoms of vaginismus and dyspareunia.

In summary, gender identity and gender role behaviors appear to be clearly female across TS samples. However, the sexual behavior data suggest that sexual interest and/or functioning are somewhat low among TS women. It is not clear if this sexual inhibition is secondary to genetic/hormonal factors or to social discomfort, given the knowledge of infertility and short stature with which these women are also coping.

## PSYCHOPATHOLOGY

A second major area of psychosocial investigation has focused on determining whether or not this genetic syndrome is associated with an increased risk for psychopathology. Case reports of TS women have described significant psychopathology, including schizophrenia and depression (Raft, Spencer, & Toomey, 1976; Sabbath et al., 1961), but a uniform diagnostic picture has not been identified. At least 11 cases of anorexia nervosa in TS women have been documented (Darby, Garfinkel, Vale, Kirwan, & Brown, 1981; Halmi & DeBault; 1974; Kron, Katz, Gorzynski, & Weiner, 1977; Walinder & Mellbin, 1977). However, the relationship with anorexia nervosa is still rare in the Turner population and may represent a chance co-occurrence (Darby et al., 1981).

Since 1955, at least sixteen studies have assessed the psychiatric history and mental status of samples of TS women. As can be seen in Table 4.2, the early studies on groups of women with Turner syndrome did not describe a consistent pattern of psychiatric problems (Garron & Vander Stoep, 1969; Hampson et al., 1955; Kihlbom, 1969; Shaffer, 1963). In a more detailed study of psychopathology, Money and Mittenthal (1970) reviewed clinical interviews of 68 patients and their parents and investigated the relationship between psychosocial functioning of the girls, family adjustment, and parenting style. Only three of the 68 girls and women evaluated were rated as having "severe" psychopathology, and six others were seen as evidencing "mild" psychopathology. While unrelated to either the number of physical anomalies or to pubertal status, psychopathology was seen most frequently in girls whose parents had overt psychopathology, were rejecting of their daughters, or, to a lesser extent, were overprotective.

Three other large-scale studies of the psychiatric and psychosocial functioning of women with Turner syndrome have been conducted in Scandinavia (Nielsen & Sillesen, 1981; Nielsen et al., 1977; Taipale, 1979). Nielsen and Sillesen initially assessed the prevalence of psychiatric disorders in TS women by screening women in psychiatric hospitals throughout Denmark. Eight women with Turner syndrome were identified via this procedure, representing a lower prevalence than would have been expected from population frequency data projections. Next, 45 TS females between the ages of 7 and 39 years were evaluated via clinical interviews and review of hospital records. Similar data were collected for two control groups, 45 natural sisters of the Turner subjects, and a group of 15 women with short stature and primary amenorrhea, but a normal chromosomal complement. A very low prevalence of psychopathology was reported for all three groups with Turner syndrome.

Using the system outlined by Money and Mittenthal (1970), only one of the TS women was rated as having severe psychopathology, while four were placed in the mild psychopathology category. Nielsen and colleagues (1977) also found a significant relationship between the identified patients' psychopathology and family environmental factors. They reported that while all five of those seen as having significant difficulties came from stressful family environments, this held true for only nine of the 40 girls and women who were without significant psychopathology. The researchers judged that four of these five were reared in overprotective environments.

In the second Scandinavian study, conducted in Finland, Taipale (1979) evaluated the psychiatric and social adjustment of 49 TS girls and young women to assess the impact of hormone replacement. Four of these young women were found to have significant emotional difficulties that Taipale (1979) determined to be associated with

**TABLE 4.2**
**Studies Including Some Assessment of Psychopathology and/or Personality in TS Women**

| | Study | Sample Size | Age Ranges | Country | Controls | Measures* | Results |
|---|---|---|---|---|---|---|---|
| 1. | Hampson, Hampson, & Money (1955) | 11 | 9-27 | U.S. | None | Interviews Thematic Apperception Test (TAT) | No major psychopathology 1 pt w/mild maladjustment 9 of 11 seem as overly inhibited and lacking initiative |
| 2. | Sabbath et al. (1966) | 7 | 14-20 | U.S. | None | Interviews | 1 schizophrenic; "Overcompliant, passive, and inhibited manner" found especially in those under 55 inches in height; Attitudes of mothers significantly influence on girls' adjustment |
| 3. | Shaffer (1963) | 13 | 13-23 | U.S. | Test Norms | MMPI, Guilford Zimmerman Temperament Survey (GZTS) | MMPI: TS significantly lower than normative sample on Hypomania scale GZTS:TS significantly lower than norms on General Activity Energy Scale and Masculinity-Femininity; higher on Personal Relations/ Cooperativeness |

*Only measures relevant to this table are listed.

**TABLE 4.2**
**Studies Including Some Assessment of Psychopathology and/or Personality in TS Women (continued)**

| | Study | Sample Size | Age Ranges | Country | Controls | Measures* | Results |
|---|---|---|---|---|---|---|---|
| 4. | Kihlbom (1969) | 11 | 6-23 | Sweden | None | Interviews TAT (N=5) Rorschach (N=5) | No psychoses 1 Pt. w/Anorexia Nervosa 5 Pts. w/neurotic problems All TS pts seen as immature; non-aggressive and compliant personality styles |
| 5. | Money & Mittenthal (1970) | 73 | 10-Adult | U.S. | None | Interviews Chart Review | Psychopathology: 3 Severe, 6 mild; Personality: majority of pts seen as "phlegmatic, complaint with high tolerance for adversity" Parental attitudes associated with girls' adjustment |
| 6. | Rothchild & Owens (1972) | 11 | 9-29 | U.S. | None | Interviews TAT Vineland Soc Maturity Inv. Sentence Compl. Rorschach, Dreese-Mooney Interest Inventory | Immaturity, muting of affect noted in younger Ss; older Ss with hormone tx seen as more mature but with deficits in heterosexual activities, and understanding, empathy, and foresight |

**TABLE 4.2**
**Studies Including Some Assessment of Psychopathology and/or Personality in TS Women (continued)**

| | Study | Sample Size | Age Ranges | Country | Controls | Measures* | Results |
|---|---|---|---|---|---|---|---|
| 7. | Perheentupa et al. (1974) | 26 | 9-17 | Finland | None | Interviews | Good social/emotional development 9-12 yrs; Immaturity, regression, isolation if pubertal changes are delayed |
| 8. | Nielsen, Nyborg, & Dahl (1977) | 45 | 7-39 | Denmark | 15 short stature Amenorrheic normal Records; karyotype women (SS); 46 sisters of TS women ± 5 yrs | Psychiatric Interviews Maudsley Personality Inventory (MPI) | Psychopathology: 1 severe (Anorexia Nervosa) 4 Mile 4/5 with psychopathology raised in overprotective homes MPI: TS significantly lower than of controls on neuroticism scale. Delayed social and sexual behavior in TS and SS controls in contrast to sisters |
| 9. | Taipale (1979) | 49 | 9-22 | Finland | None | Interview Draw-A-Tree Draw-A-Person (DAP) Rorschach | Psychopathology: 4 severe, associated with delayed hormone therapy |

**TABLE 4.2**
**Studies Including Some Assessment of Psychopathology and/or Personality in TS Women (continued)**

| | Study | Sample Size | Age Ranges | Country | Controls | Measures* | Results |
|---|---|---|---|---|---|---|---|
| 10. | Nielsen & Sillesen (1981) | 103 (82 complete evaluations) | 7-24 | Denmark | None | Interviews Psychiatric Records | Psychopathology: None 23/86 behavior problems: shyness, sensitivity, insecurity 50/86 emotionally immature Behavior problems were related to parenting styles |
| 11. | Chen, Faigenbaum, & Weiss (1981) | 24 | 9 mos - 18 yrs | U.S. | None | Pediatric Interviews | Psychopathology: None Good social skills |
| 12. | Peri & Molinari (1983) | 48 | 10-14 | Italy | None | Interviews Rorschach, (DAP) Draw-A-Tree Pyramid of Color | Psychopathology: None Introverted with difficulty speaking about inner conflicts, including anxiety as move into adolescence |
| 13. | Sonis et al. (1983) | 16 | 6-16 | U.S. | 16 Matched f/age, sex, race, and SES | Child Behavior Checklist (CBCL) | CBCL: TS increase controls hyperactive/aggressive and Schizoid/Anxious/Obsessive Scales |
| 14. | Rovet (1986) | 29 | 7-16 | Canada | | CBCL Middle Childhood Temperament Survey (MCTS) | CBCL: Increased hyperactivity for younger Ss ≤11 MCTS: Easily distracted, non-persistent style |

delayed hormone replacement treatment and subsequent lack of pubertal development.

Many studies of genetic anomalies are troubled by sampling bias and/or a tendency to include subjects within wide age ranges to achieve an adequate sample size. Therefore, in the third Scandinavian study, an attempt was made to follow up and evaluate every girl with Turner syndrome born between 1955 and 1966, thereby providing the first study of an unselected sample (Nielsen & Sillesen, 1981). One hundred and three girls were identified in all, and 86 of them along with their parents participated in a psychiatric interview. Further data on psychosocial functioning were obtained from physician and school records. None of the girls had a history of psychiatric hospitalization or treatment. The majority of the sample was described as functioning well, having only minor difficulties with immaturity. Twenty-three girls were identified in the clinical study as having behavior problems centering primarily on difficulties with shyness, sensitivity, and insecurity. Five girls who were judged as having some "mental" problems were the same five in the sample who had below-average intellectual abilities. Parents described the TS girls as easier to get along with and more conscientious than their sisters. Although few girls were felt to have significant behavior problems, 58% were seen as below average in emotional maturity. As in the earlier studies by Money and Mittenthal (1970) and Nielsen and colleagues (1977), this was strongly related to parental style. Behavior problems and immaturity were found more commonly in girls raised by overly restrictive, indulgent, or protective parents.

In our study of the adult adjustment of TS women (McCauley, Sybert, & Ehrhardt, 1986), six of 27 women who completed a detailed psychiatric interview met criteria for a major psychiatric diagnosis. One woman had a history of Obsessive Compulsive Disorder, with repeated handwashing, during early adolescence and had been in psychotherapy for approximately two years at that time. The most common problem in this sample was depression with related anxiety symptoms. However, since depression is a problem that affects many adult women, the lack of a control group in this study makes it difficult to evaluate these findings. However, additional data were collected using the Tennessee Self Concept Scale (TSCS), a well standardized, multidimensional self-report scale. In contrast to the normative sample, the TS women exhibited significantly less positive self-esteem on multiple subscales of the TSCS. Furthermore, on some of the scales that screen for psychopathology, the TS women had scores that were more similar to those of the psychiatric patients than to the normative, nonclinical sample. These data suggest that although TS women may not have an increased prevalence of major psychiatric disorders, they do have difficulties coping with their situation as reflected in low self-esteem.

Two recent studies focusing on girls and young adolescents with TS have reported not only a low prevalence of major psychopathology, but also findings suggestive of significant conflict.

Peri and Molinari (1983) used projective testing and clinical interviews to assess the functioning of 48 Turner syndrome patients (ages 10 to 14 years). While no severe psychopathology was identified, the girls were seen as introverted with difficulty discussing their inner feelings. Rorschach testing was interpreted as revealing anxiety and a sense of insecurity. Using a very different approach, Sonis and colleagues (1983) compared parental report on the Child Behavior Checklist (CBCL) for 16 TS girls, ages 6 to 16, with similar parental report data collected from a control group matched for age, race, and socioeconomic status. The TS girls were rated significantly higher than controls on the Hyperactive/Aggressive and the Schizoid/Anxious/Obsessive scale. Problems with ability to concentrate and heightened perceived activity level were more prominent in the younger age group (6-11), with more social withdrawal and depressed/anxious behavior appearing in the older girls.

The studies reviewed thus far document consistently a low prevalence of psychopathology in women with Turner syndrome. Each sample of women may include a small subset with major psychiatric diagnoses, but there is not a clear diagnostic profile that characterizes TS women. Furthermore, those individuals with greatest impairment are frequently described as coming from dysfunctional families or as having parents who themselves were unable to cope successfully with their daughter's diagnosis. On the basis of the 470 to 480 TS females represented in the studies reviewed here, one can estimate a prevalence rate of 2-6% for major psychopathology. This is similar to the prevalence rates (3.7-7.6%) reported in three epidemiological samples of adults conducted in the United States (Robins et al., 1984). The data appear consistent enough to conclude that Turner syndrome is not associated with an increased risk for a major psychiatric disorder.

## PERSONALITY STYLE AND SOCIAL ADJUSTMENT

Although psychiatric disorders were not common in the studies reviewed above, problems in overall adjustment characterized by immaturity and overcompliance were frequently described (see Table 4.2). While most of these studies presented summaries of clinical evaluations, Shaffer's (1963) investigation included standardized assessment devices: the Minnesota Multiphasic Personality Inventory (MMPI) and the Guilford-Zimmerman Temperament Survey (GZTS). Data from these measures were obtained for only 13 subjects. The adolescent girls with Turner syndrome differed from normative groups in that their scores were significantly lower on the Hypomanic scale of the MMPI suggesting a low energy, nonreactive personality style. In contrast to normative samples of women, the TS sample scored significantly lower on the Masculinity-Femininity (indicating stronger femininity scores) and Energy scales of the GZTS and higher on the scale reflecting cooperation. Because of these differences and the lack of variability in the response patterns of the 13 TS women, Shaffer suggested that lowered energy and low impulsivity/reactivity might be traits associated with this genetic syndrome. Money and Mittenthal (1970) also observed a similarity in presentation across their sample, leading them to suggest a common personality style characterized by subdued emotional arousal with an unusually "high tolerance for adversity," unassertiveness, and overcompliance.

A comprehensive study by Nielsen et al. (1977) included an assessment of social functioning as well as psychopathology. These researchers found that in relation to educational attainment and job status, TS women did not differ significantly from their sisters. However, both the TS women and the short-stature control subjects were less likely than the sister group to live independently from their parents; furthermore, fewer women in these two groups were married or sexually active. These results suggest that short stature and delayed sexual development are key factors influencing psychosocial development quite independently from the sex chromosomal anomaly.

An alternative causal model was introduced in an additional component of the Nielsen et al. (1977), study. Baekgaard, Nyborg, and Nielsen (1978) investigated the Maudsley Personality Inventory (MPI) responses of 31 TS women in comparison to three control groups: 16 TS sisters, 9 short-stature, amenorrheic women with no genetic abnormality, and 19 nurses. The two personality dimensions assessed by this scale are neuroticism and extraversion. The response patterns of all three control groups were comparable to those of the normative samples on both dimensions and did not differ significantly from each other. In contrast, the scores of the TS sample fell significantly below those of the control groups on the neuroticism scale and were elevated on the extraversion dimension. Furthermore, subanalyses indicated that these differences reflected the response pattern of the subgroup (N = 13) of TS women with 45,X karyotypes and did not hold for those with other karyotypes when analyzed separately. These results were interpreted as suggesting that the loss of the second sex chromosome leads to a personality configuration characterized by limited emotional arousal and unusually high stress tolerance. This personality pattern was seen as directly related to the extent of the loss of the second X chromosome.

A somewhat different perspective emerges from a Finnish study. Taipale (1979) conducted an in-depth psychiatric and psychosocial evaluation of 49 girls and adolescents with Turner syndrome to assess the impact of hormone replacement on adolescent development. Taipale found no uniformity in personality structure among her subjects and described them as anxious and upset about their condition. There was no evidence of the kind of decreased emotional arousal or ready acceptance of misfortune described by Money and Mittenthal (1970). Furthermore, Taipale's subjects expressed having more difficulty maintaining good peer relationships than had been reported previously. In this sample, psychological distress was much more severe and prevalent in those young women who had experienced a significant delay in the onset of pubertal changes. Because of the profound effects of delayed puberty on emotional development and sense of well-being, Taipale and colleagues (Perheentupa et al., 1974; Taipale, 1979) concluded that hormone replacement therapy should be initiated in a time frame that allows the TS girl to mature with her peers. These researchers also interpreted the immaturity and social withdrawal found in their subjects as a behavioral adaptation to being short-statured and a late maturer rather than an offshoot of the sex chromosome anomaly.

Rothchild and Owens (1972) used periodic clinical interviews over three years to follow the developmental course of two groups of TS patients, girls 9 to 14 years old, and a group of older (17-29 years), late-diagnosed women. They found immaturity in relation to adolescent psychological growth following the onset of hormone replacement therapy. The older group took a more active stance in pursuing information regarding their condition, but were still immature in terms of social and sexual development. Rothchild and Owens noted that even in light of successful academic and professional attainment, these young women experienced difficulty in social interactions and in "attentiveness to common nuances of feeling."

Recent studies done with North American samples of child and adolescent subjects also document behavior or adjustment problems. As mentioned above, Sonis et al. (1983) found problems in the ability to concentrate and heightened perceived activity level in their sample of younger (6-11 years) TS girls, but more social withdrawal and depressed/anxious behavior in the older TS girls. Sonis and colleagues attribute the attentional problems found in the younger TS subjects to immaturity in central nervous system development. Similar findings were reported by Rovet (1986) when she administered the Middle Childhood Temperament Questionnaire (MCTQ) and Child Behavior Checklist (CBCL) to the parents of girls with Turner syndrome. Data from the MCTQ, based on 18 girls age 11 years or younger, indicated an ongoing but easily distracted, nonpersistent style. The CBCL data were collected for two groups of TS girls, 14 of whom were age 11 years or younger and 11 who were 12 to 16 years old. In the younger group, an increased incidence of hyperactivity was again found, while anxiety was more apparent in the older subjects.

The findings from this latter group of studies are not consistent with earlier reports of a low arousal, low energy personality style. In fact the only theme common to all studies is immaturity in psychosocial development. Both Rovet (1986) and Sonis et al. (1983) concluded that the behavioral patterns they found were secondary to CNS immaturity, which is in turn a result of the chromosomal anomaly. In contrast, Taipale and colleagues (Perheentupa et al., 1974; Taipale, 1979) view these characteristics as simply behavioral adaptations to being short statured and sexually immature.

Therefore, we are left with questions about whether a distinct personality or behavioral style is associated with this syndrome, and if so, whether it is part of the syndrome itself or the result of the physical immaturity the syndrome causes. Furthermore, if a characteristic behavioral style does exist, what is the mechanism by which the genetic anomaly influences behavior?

## INTERACTION BETWEEN SOCIAL AND COGNITIVE BEHAVIOR

In an attempt to address the question raised above regarding behavioral patterns and the origins of such patterns, we conducted a detailed assessment of the cognitive and psychosocial functioning of a group of TS girls and a group of like-aged girls with familial short stature (McCauley, Ito, & Kay, 1986; McCauley, Kay, Ito, & Treder, 1987). The two groups were comparable in age, height, weight, overall Verbal IQ, and family socioeconomic status. Psychosocial assessment included use of the Child Behavior Checklist (CBCL), the Piers-Harris Children's Self-Concept Scale, and child, parent, and teacher reports of social behavior. Cognitive assessment consisted of age-appropriate Wechsler intelligence scales, Embedded Figures (Witkin, Oltman, Raskin, & Karp, 1971) scales, and the Developmental Test of Visual-Motor Integration (Beery, 1967). The purpose of this study was twofold. The first was to explore the hypothesis that the behavior patterns associated with Turner syndrome are adaptations to being short statured and sexually immature. The second purpose was to investigate the relationship between the psychosocial functioning of girls with Turner syndrome and the cognitive deficits that have been documented in these individuals.

The results of the social-emotional assessment revealed significant group differences in social functioning, behavioral problems, and self-esteem. The subjects with Turner syndrome were rated by self and others (mothers and teachers) as less socially adept than the controls across the social competency and self-concept measures. The CBCL data revealed significant group differences for the Immature-Hyperactive and the Depressed-Withdrawn scales. Although the girls with Turner syndrome were rated as having more problems in both areas than the short-stature controls, scores on both scales were still within the normal range (55-70 $T$ scores). However, the Immature-Hyperactive Mean $T$ score was very close to the range suggestive of significant psychopathology ($T = 69.06$).

As with the CBCL and the other measures of social adjustment, the TS girls were consistently depicted as having difficulties with peer relationships. These problems were characterized as frequent teasing and not being liked or included by other children. Social and emotional difficulties were found within the sample of girls with Turner syndrome, but not within the short-stature control group. The short-stature group, although teased about their size, did not show the same prevalence of behavior or peer problems and had positive self-image scores. Consequently, short stature alone cannot account for the social and emotional difficulties found in the Turner syndrome group.

A number of other factors could underlie these difficulties. First, is that of delayed sexual maturation. We were able to control for short stature, but not timing of pubertal changes. However, delayed pubertal timing does not appear to completely account for the social deficits found in our sample. All but the two oldest subjects had started receiving hormone treatment and thus began pubertal development by age 13. Moreover, one of the two older girls had some spontaneous sexual development. A second important factor is the role of the physical anomalies that sometimes accompany Turner syndrome, such as webbing of the neck and digital defects. These anomalies affect the physical appearance of TS girls and thus may lower self-confidence or cause others to react negatively to these girls. In the sample described here, the number of external physical anomalies was not correlated with parent or teacher ratings of social competence, but did correlate with subjects' ratings on the Piers-Harris Self-Concept Scale. A greater number of physical anomalies was significantly correlated with a lower self-image score on the physical appearance and anxiety subscales.

Another critical factor could be differences in brain development and maturation. Research on the cognitive development of Turner syndrome patients has consistently documented visuo-spatial problem-solving difficulties suggestive of right hemisphere deficits (Alexander, Ehrhardt, & Money, 1966; Money, 1973; Silbert, Wolff, & Lilenthal, 1977), and has also provided evidence of more global deficits (Pennington et al., 1985; Waber, 1979). These deficits have been attributed to differences in the process of brain maturation secondary to the lack of prenatal and pubertal ovarian hormone activity. The cognitive assessment included in our study of girls with Turner syndrome was done to determine whether similar visual-motor problems were present in our sample. On the cognitive tests, significant group differences were found on the Arithmetic, Digit Span, Picture Completion and Object Assembly subtests of the Wechsler Intelligence Scales and on the Developmental Test of Visual-Motor Integration. In all cases, subjects with Turner syndrome performed less well than short-stature controls. These results indicated spatial and attentional difficulties that were consistent with the pattern of previous findings of cognitive deficits in TS women.

Since there was a growing body of data suggesting both cognitive and social-behavioral problems, we wondered if there could be a relationship between the two. This led to the second component of our study comparing TS girls to those with familial short stature (McCauley et al., 1987). In addition to documenting social and cognitive problems, a task requiring interpretation of facial affect (Kay, 1983) was included in an effort to investigate the relationship between social and cognitive functioning.

We hypothesized that the ability to accurately discriminate facial cues of affect could be a "bridge" linking the cognitive and psychosocial problems known to exist in girls with Turner syndrome. Since parents described TS girls as having trouble picking up on the social context and being oblivious to how others were reacting to them, we speculated that they might be having difficulty interpreting subtle social cues, such as those represented in facial expressions. Our hypothesis was based on the documentation of cognitive problems in females with Turner syndrome suggestive of right hemisphere involvement. Facial and affective processing is also thought to be a right hemisphere function (Hilliard, 1973; Landis, Assal, & Perret, 1979; Leehey & Cahn, 1979; Ley & Bryden, 1979; Tucker, 1981).

The girls with Turner syndrome were less accurate at correctly reading facial affect than the control group (McCauley et al., 1987). Additional analyses indicated that these differences held up even when spatial processing (as reflected in Performance IQ scores) and attentional/memory problems (as reflected in the Wechsler attentional factor) were controlled. Thus, the results suggest a difficulty with interpretation of facial affect that is independent of visuo-spatial problem-solving skills. The two skills may be independent of each other, but both might be linked to an underlying right hemisphere function.

What does this tell us about the question of whether or not there is a personality configuration associated with the deletion of the second X chromosome? It may be that what has been interpreted as a personality pattern is the reflection of cognitively-based perceptual difficulties that make it hard for individuals with Turner syndrome to pick up on more subtle, nonverbal social cues. This deficit, along with the social immaturity that characterizes these girls, could give rise to the impression that they are emotionally inert; however, the more recent data suggest a more complex picture of underlying behavior and self-esteem problems.

The social immaturity appears to be the result of at least two major factors. First, a number of the studies described above (Money & Mittenthal, 1970; Nielsen & Sillesen, 1981; Nielsen et al., 1977) have noted that social immaturity was highly related to family environment and parental style. There is a very strong tendency for parents and friends of TS girls to treat them according to their size, and consequently, to have lowered expectations regarding social maturity. This frequently extends beyond the family to school and peer situations. Furthermore, these girls clearly show a lag in terms of their psychosexual development, which is ameliorated to some extent by timely hormone replacement. Hormone replacement can only mimic the natural cycle of fluctuating hormonal levels and does not provide the gradual hormonal changes that are the precursors of pubertal changes in girls with functioning ovaries.

## SUMMARY AND CONCLUSIONS

The results of the studies to date provide consistent support for the fact that the majority of individuals with Turner syndrome are functioning as very productive individuals. There appear to be some commonalities in behavioral patterns, such as an increased tendency for peer and attentional problems in childhood, followed by delayed psychosexual maturation and ongoing difficulties with interpretation of social cues. These difficulties seem to be related to delays in brain maturation and may resolve as the TS individual moves into adult life. These people on the whole are coping fairly well and appear to handle their condition and its limitations without a great deal of complaint, but it does not seem warranted to minimize their coping by ascribing it to a characteristic personality style. It would appear more beneficial to identify potential trouble spots, such as peer relationships, age-appropriate maturation, and attention to social cues, thereby helping TS girls and their families to address these problems throughout the developmental process, rather than glibly reassuring parents that TS girls cope well.

To more accurately investigate the question of personality style, further assessment of personality dimensions based on larger samples is needed. Samples large enough to tease out differences among subtypes would be ideal, as well as attempts to control for variations in cognitive functioning. This may well require cooperative research efforts across clinical sites. Furthermore, it will be important to evaluate the impact of early diagnosis and to have more open discussion of the diagnosis and the implications.

## ACKNOWLEDGMENTS

The author wishes to express appreciation to Betty Compton for her help in preparing this manuscript.

## REFERENCES

Alexander, D., Ehrhardt, A. A., & Money, J. (1966). Defective figure drawing, geometric and human in Turner's syndrome. Journal of Nervous and Mental Disease. 42, 161-167.

Baekgaard, W., Nyborg, H., & Nielsen, J. (1978). Neuroticism and extroversion in Turner's syndrome. Journal of Abnormal Psychology, 87, 583-586.

Beery, K. E. (1967). Developmental Test of Visual Motor Integration. Administration and Scoring Manual. Chicago, IL: Follett Publishing Co.

Chen, M., Faigenbaum, D., & Weiss, H. (1981). Psychosocial aspects of patients with the Ullrich-Turner syndrome. American Journal of Medical Genetics, 8, 191-203.

Cohen, H. (1962). Physiological test findings in adolescents having ovarian dysgenesis. Psychosomatic Medicine, 24, 249-256.

Darby, P. L., Garfinkel, P. E., Vale, J. M., Kirwan, P. J., & Brown, G. M. (1981). Anorexia nervosa and Turner syndrome: Cause or coincidence? Psychological Medicine, 11, 141-145.

Ehrhardt, A. A., Greenberg, N., & Money, J. (1970). Female gender identity and absence of fetal hormones: Turner's syndrome. Johns Hopkins Medical Journal, 126, 237-248.

Halmi, K. A., & DeBault, L. E. (1974). Gonosomal aneuploidy in anorexia nervosa. American Journal of Genetics, 26, 195-198.

Hampson, J. L., Hampson, J. C., & Money, J. (1955). The syndrome of gonadal agenesis (ovarian agenesis) and male chromosomal pattern in girls and women: Psychologic studies. Bulletin of Johns Hopkins Hospital, 97, 207-226.

Hilliard, R. D. (1973). Hemispheric laterality effects on the facial recognition task in normal subjects. Cortex, 9, 246-258.

Kay, T. (1983). Individual differences in children's abilities to discriminate positive and negative effect. Unpublished doctoral dissertation. Emory University, Atlanta, GA.

Kihlbom, M. (1969). Psychopathology of Turner's syndrome. Acta Paedopsychiatrica, 36, 75-81.

Kron, T., Katz, J. L., Gorzynski, G., & Weiner, H. (1977). Anorexia nervosa and gonadal dysgenesis. Archives of General Psychiatry, 34, 332-335.

Landis, T., Assal, G., & Perret, E. (1979). Opposite cerebral hemispheric superiorities for visual associative processing of emotional facial expressions and objects. Nature, 278, 739-740.

Leehey, S., & Cahn, A. (1979). Lateral asymmetries in the recognition of words, familiar faces, and unfamiliar faces. Neuropsychologia, 17, 619-635.

Ley, R. G., & Bryden, M. P. (1979). Hemispheric differences in processing emotions and faces. Brain and Language, 7, 127-138.

McCauley, E., Ito, J., & Kay, T. (1986). Psychosocial functioning in girls with the Turner syndrome and short stature. Journal of the American Academy of Child Psychiatry, 25, 105-112.

McCauley, E., Kay, T., Ito, J., & Treder, R. (1987). The Turner syndrome: Cognitive deficits, affective discrimination and behavior problems. Child Development, 58, 464-473.

McCauley, E., Sybert, V., & Ehrhardt, A. A. (1986). Psychosocial adjustment of adult women with Turner syndrome. Clinical Genetics, 29, 284-290.

Money, J. (1973) Turner's syndrome and parietal lobe functions. Cortex, 9, 385-393.

Money, J., & Mittenthal, S. (1970). Lack of personality pathology in Turner syndrome: Relations to cytogenetics, hormones and physique. Behavior Genetics, 1, 43-56.

Nielsen, J., Nyborg, H., & Dahl, G. (1977). Turner's syndrome: A psychiatric-psychological study of 45 women with Turner's syndrome, compared with their sisters and women with normal karyotypes, growth retardation and primary amenorrhea. Acta Jutlandica, XLV, Medicine Series 21, Arhus.

Nielsen, J., & Sillesen, I. (1981). Turner's syndrome in 115 Danish girls born between 1955 and 1966. Acta Jutlandica, LIV, Medicine Series 22, Arhus.

Pennington, B. F., Heaton, R. K., Karzmark, P., Pendleton, M. G., Lehman, R., & Shucard, D. W. (1985). The neuropsychological phenotype in Turner syndrome. Cortex, 21, 391-404.

Perheentupa, J., Lenko, H. L., Nevalainen, I., Nittymaki, M., Soderholm, A., & Taipale, V. (1974). Hormone therapy in Turner's syndrome: Growth and psychological aspects. Pediatric XIV. Growth and Developmental Endocrinology, 5, 121-127.

Peri, G., & Molinari, E. (1983). Psychological aspects in gonadal dysgenesis. Acta Medical Auxology, 15, 75-84.

Pollock, G. H. (1955). The psychologic response to estrogenic therapy in Turner's syndrome (ovarian agenesis). Journal of Nervous and Mental Disorder, 121, 420-422.

Raft, D., Spencer, R. F., & Toomey, T. C. (1976). Ambiguity of gender identity fantasies and aspects of normality and pathology in hypopituitary dwarfism and Turner's syndrome: Three cases. Journal of Sex Research, 12, 161-172.

Robins, L. N., Helzer, J. E., Weissman, M. M., Orvaschel, H., Gruenberg, E., Burke, J. D., & Regier, D. A. (1984). Lifetime prevalence of specific psychiatric disorders in three sites. Archives of General Psychiatry, 41, 949-958.

Rothchild, E., & Owens, R. P. (1972). Adolescent girls who lack functioning ovaries. Journal of the American Academy of Child Psychiatry, 11, 88-113.

Rovet, J. (1986). Processing deficits in 45.X females. Paper presented at the American Association for the Advancement of Science Annual meeting, Philadelphia, PA.

Sabbath, J. C., Morris, T. A., Menzer-Benaron, D., & Sturgis, S. H. (1961). Psychiatric observation in the adolescent girls lacking ovarian function. Psychosomatic Medicine, 23, 224-231.

Shaffer, J. W. (1963). Masculinity-femininity and other personality traits in gonadal aplasia (Turner's syndrome). In H. C. Beigel (Ed.), Advances in sex research (pp. 219-232). New York: P. P. Hoeber & Sons, Inc.

Silbert, A., Wolff, P. H., & Lilienthal, J. (1977). Spatial and temporal processing in patients with Turner's syndrome. Behavior Genetics, 7, 11-21.

Sonis, W. A., Levine-Ross, J., Blue, J., Cutler, G. B., Loriaux, P. L., & Klein, R. P. (1983, October). Hyperactivity of Turner's Syndrome. Paper presented at American Academy of Child Psychiatry meeting, San Francisco, CA.

Taipale, V. (1979). Adolescence in Turner's syndrome. Monograph from Children's Hospital, Univerity of Helsinki, Helsinki, Finland.

Theilgaard, A. (1972). Cognitive style and gender role in persons with sex chromosome aberrations. Danish Medical Bulletin, 19 276-286.

Tucker, D. M. (1981). Lateral brain function, emotion, and conceptualization. Psychological Bulletin, 89, 19-46.

Waber, D. P. (1979). Neuropsychological aspects of Turner syndrome. Developmental Medicine and Child Neurology, 21, 58-70.

Walinder, J., & Mellbin, G. (1977). Karyotyping of women with anorexia nervosa. British Journal of Psychiatry, 13, 48-49.

Witkin, H. A., Oltman, P. K., Raskin, E., & Karp, S. A. (1971). A Manual for the Embedded Figures Tests. Palo Alto, CA: Consulting Psychologists Press, Inc.

# 5

# Sex Hormones, Brain Development, and Spatio-Perceptual Strategies in Turner Syndrome

Women with Turner syndrome (TS) typically encounter severe difficulties when asked to solve various spatial ability tasks. Despite numerous attempts, agreement has not been reached about the specific nature or cause(s) of their spatial deficits. In this chapter some aspects of this dilemma will be illustrated, and possible solutions proposed. The first half of the chapter involves a detailed examination of the spatio-perceptual strategies used by TS women in the Rod-and-Frame Test (RFT), a measure of field dependence-independence. In the second half, I discuss specific hormonal factors that may explain not only the spatio-perceptual deficits exhibited by TS women in the RFT, but also a number of other phenomena concerning sex-related variations in spatial skills.

## FIELD DEPENDENCE-INDEPENDENCE

In a series of studies, my colleagues and I demonstrated clearly that TS women score highly field dependent on tasks like the RFT and the Embedded-Figures Test (Nielsen, Nyborg, & Dahl, 1977; Nyborg & Nielsen, 1981a). How should this finding be interpreted? Over a number of years, the creators of the field dependence-independence dimension revised their description of field dependence (Asch & Witkin, 1948; Witkin, 1964; Witkin & Asch, 1948; Witkin, Dyk, Faterson, Goodenough & Karp, 1962/1974; Witkin & Goodenough, 1977a; 1977b; 1981; Witkin et al., 1954; Witkin & Oltman, 1967). The traditional way of scoring performance in the RFT was criticized, and new methods suggested (Arbuthnot, 1972; Gruen, 1957; Haller, 1981; Haller & Edgington, 1982a, 1982b; Lester, 1968, 1971; McGarvey, Maruyama, & Miller, 1977; Nyborg, 1977). Because the old scoring method was shown to confound a number of variables, it became rather problematic to determine the meaning of being classified as field dependent.

To find out more about the meaning of field dependence in the RFT, I began a series of conceptual and methodological studies (Nyborg, 1971a, 1971b, 1972, 1974a, 1974b; Nyborg & Isaksen, 1974). These studies yielded an alternative interpretation of task demands in the RFT along with a new method for scoring performance in this task. Subsequently, it was shown that the new scoring method could be modified to study spatio-perceptual strategies in the RFT (Nyborg, 1977). The results of this series of studies provide the basis for the present attempt to discover more about what makes the RFT such a difficult task for many TS women.

The following sections contain brief descriptions of the RFT procedure and content analyses of the task (see Nyborg, 1977 for more details). The original RFT Turner data from the Nielsen et al. (1977) study are subjected to a developmental reanalysis, and the outcome is compared with preliminary norms for RFT scores. The reanalysis concentrates on mapping age differences in the two major response parameters of the RFT. The two response-parameter values are combined in a subsequent analysis to indicate the type of spatio-perceptual strategy used by an individual in trying to cope with the experimentally-induced perceptual conflict. Finally, age differences in the spatio-perceptual strategies of TS women are compared to those of unaffected schoolgirls.

## Subjects

Data from 45 women with Turner syndrome (Turner, 1938) were available. Twenty-one had a 45,X karyotype, while the other 24 showed either a loss of part of the second X chromosome or mosaicism (for details, see Nielsen et al., 1977). Fifty percent lived in Copenhagen, while the rest lived in smaller towns or rural districts of Denmark. Their ages ranged from 7.1 to 38 years with a mean of 20.9. They were arbitrarily divided into five age groups: (1) younger than 15, (2) 15-19, (3) 20-24, (4) 25-29, and (5) older than 30. The mean ages for the five groups were 12.11, 17.7, 22.6, 27.0, and 34.9 years, respectively. About half of the subjects had received cyclic estrogen-gestagen therapy for various periods of time, and duration of treatment has been shown to be related to the expression of spatial ability (Nyborg & Nielsen, 1981a).

The control group consisted of Danish schoolgirls aged 8, 10, 12, 14, and 16 years (plus or minus one month) who volunteered to be studied with the RFT. These girls were similar to the Turner sample in SES.

## Instrument and Procedure

A transportable RFT apparatus was used with the subject placing her head at one end of the box so that her view was restricted to the inside. A square frame with a movable rod inside was visible at the other end of the box. The rod and the frame were then tilted initially 28 degrees to the right or to the left of gravitational vertical in all possible combinations over four trials. This was repeated once to obtain a total of eight measures. The subject's task was to adjust the rod to a physically vertical position within the stationarily tilted frame.

Content analysis. A content analysis of the RFT has shown that the following variables account for most of the variance in performance: (1) the potentially misleading frame tilt information (frame effect: *phi*), (2) the effect of the rod-starting-position (rod effect: *rho*), (3) some subjects' permanently deviating perception of the physical upright (subject error: *mu*), (4) the subject's degree of response consistency to identical rod-frame tilt conditions (response consistency: *sigma*) (Nyborg, 1974a, 1977; Nyborg & Isaksen, 1974).[1]

Quantitative scoring method. The Witkin group recommends the unsigned deviation (USD) or absolute error scoring method for the RFT. However, the USD method has been shown to confound the four variables mentioned above. Therefore, a new method of scoring the RFT, the analytic component (AC) method, was developed (Nyborg, 1974a) and subsequently tested (Jahoda & Nielsen, 1986; Nyborg & Isaksen, 1974; O'Connor & Shaw, 1978). It was applied and developed further in the present study.

With the AC method, the direction of the deviation of the rod from gravitational vertical is recorded to calculate a signed deviation score. A record is kept of whether the rod is adjusted to the same side to which the frame is tilted or to the opposite side. The "emphasis" a subject put on the potentially misleading frame tilt information for the final adjusted position of the rod (frame effect: *phi*) is calculated in degrees from the data on signed deviation of rod setting from gravitational vertical. The tendency of a subject to adjust the rod consistently to one side of physical vertical (subject error: *mu*) is noted. The tendency of a subject to see the rod as vertical even though it is still inclined toward its originally tilted position (rod effect: *rho*) is recorded. In this way the AC method partials out the single most important spatial variables in the RFT, thereby permitting the specific sources of each subject's deviation to be traced. Each of these variables has been shown to contribute independently to a subject's overall score (Nyborg, 1974a, 1974b, 1977; Nyborg & Isaksen, 1974).

The *mu* and *rho* values were subtracted from the subject's mean overall signed deviation score. By doing so, the effect of the potentially misleading frame tilt information (i.e., the unweighted *phi*-value) is separated from other possible effects. The AC method also allows for an estimate of each subject's response consistency (*sigma*), which is a measure of the "stability" with which the subject responds to comparable conditions on initial rod-and-frame tilt. The significance of the unweighted *phi*-values can be assessed further by the AC method. This is accomplished by relating unweighted *phi* to *sigma* according to the formula $phi/sigma$ $(2\sqrt{2})$ (see Nyborg, 1974a; Nyborg & Isaksen, 1974). If the weighted *phi* value is significantly different from zero ($p < .05$), the subject is classified as "significantly emphasizing misleading frame tilt information." In other words, the weighted *phi* score provides an index of the degree of significance the subject put on the misleading frame tilt information with respect to her degree of response consistency.

Task analysis. Because the traditional USD method of scoring the RFT confounds the above-mentioned variables, it easily leads to an inadequate understanding of the stimulus situation and to ambiguity in the basic concept of field independence-dependence. For example, consider a person classified as "field independent" who must depend on the gravitational information for successful performance in the RFT. There simply is no other way to get the information needed for achieving a low error score in the RFT. The problem with the traditional field-dependence concept is that it focused almost exclusively on the impact of the visual information. Consequently, field dependence became synonymous with visual field dependence, and the aspect of gravitational information was deemphasized. Through use of the following task analysis, I attempt to redress the balance and show that the RFT is a test of spatial ability (e.g., Linn & Petersen, 1985; Nyborg, 1974a, 1977).[2]

The task analysis makes it possible to appraise which spatio-perceptual strategy a given subject adopted in order to cope with the RFT. This appraisal rests upon the following four assumptions: (1) The subject uses one of four different kinds of spatio-perceptual strategies in her attempt to solve the experimentally-induced conflict between the optic and the vestibular-somesthetic information in the RFT, (2) the *phi* and the *sigma* values provide both the necessary and sufficient information for determining which particular strategy the subject actually used, (3) the particular spatio-perceptual strategy used reflects the degree to which one is able to capitalize on vestibular-somesthetic information in the cross-modal matching with visual information, and (4) in general, as a child gets older she will be able to master still more adequate strategies. For example, it is expected that most school beginners will either respond very inconsistently to the various frame and rod tilt conditions or will be systematically and heavily influenced by the tilted visual framework. However, as they mature they become steadily more able to counter the misleading frame tilt information by gaining access to the more reliable vestibular-somesthetic information.

The spatio-perceptual strategies were calculated in the following way. Subjects were first classified in accordance with each individual's *sigma* value. Those with *sigma* above 3 degrees (an arbitrarily defined cutoff point on the response consistency-inconsistency dimension) were assigned to a category of inconsistently responding subjects. The rationale for this procedure was that regardless of the actual size of the unweighted frame tilt score, *phi*, the degree of response inconsistency was so considerable that little reliance could be put on the *phi* score; there is no way to ensure whether the *phi* value was obtained purely by chance. The high *sigma* score was taken to indicate that the subject used one or another of a variety of unsystematic D strategies (for details, see Nyborg, 1977).

The remaining systematically responding subjects (i.e., with *sigma* values below three degrees) were then classified into one of the following three categories in accordance with the size of their *phi* value:

(1) Subjects with a *phi* value above eight degrees. The combination of a low *sigma* score and a high *phi* score was taken to indicate that such subjects are consistently and significantly misled by the frame tilt information. They are said to have used a non-optimal, optically dominated C strategy.

(2) Subjects with a *phi* value between two and eight degrees. The low *sigma* score, combined with a moderately-sized *phi* score was taken to indicate that the subject consistently made a compromise between the optical and the vestibular-somesthetic information. Such subjects are said to use a near-optimal, compromise B strategy.

(3) Subjects with a *phi* score less than two degrees. A low *sigma* score combined with a low *phi* score was taken to indicate that the subject consistently was able to exploit the reliable vestibular-somesthetic information in her attempt to set the rod to a physical vertical position in the presence of potentially misleading visual tilt information. Such a subject was said to use an optimal vestibular-somesthetic A strategy.

Data analysis. By examining the RFT performance in unaffected girls age 8 to 16 years, it was possible to establish preliminary norms of age changes in spatio-perceptual strategies in the RFT from early childhood to the attainment of a stable adult level. This provided a baseline for comparing age changes in TS females.

## Results

Changes in unweighted *phi* scores. Changes with age in the ability to draw upon vestibular-somesthetic information (i.e., group mean unweighted *phi* value) in the RFT are shown in Figure 5.1. Inspection of this figure reveals that the age distribution of mean unweighted *phi* scores for TS women tends to differ from that of unaffected girls. While unaffected girls attain their optimal performance somewhere between 12 and 14 years of age, the performance of TS women peaks somewhere between 18 and 26. A slight increase in group mean unweighted *phi* is seen in both groups after the two curves reached their minimum. However, a chi-square test indicated that the age distribution of unweighted *phi* of TS women did not differ significantly from that of the unaffected girls.

Changes in response consistency, *sigma*. Figure 5.2 shows changes with age in response consistency (*sigma*) in TS women and in unaffected girls. The age distribution indicates that mean *sigma* value is at its lowest at 14 years in unaffected girls, and at a similar level around age 22 in TS women. However, there is not much variation over time in *sigma*, and no statistically significant differences between the two groups.

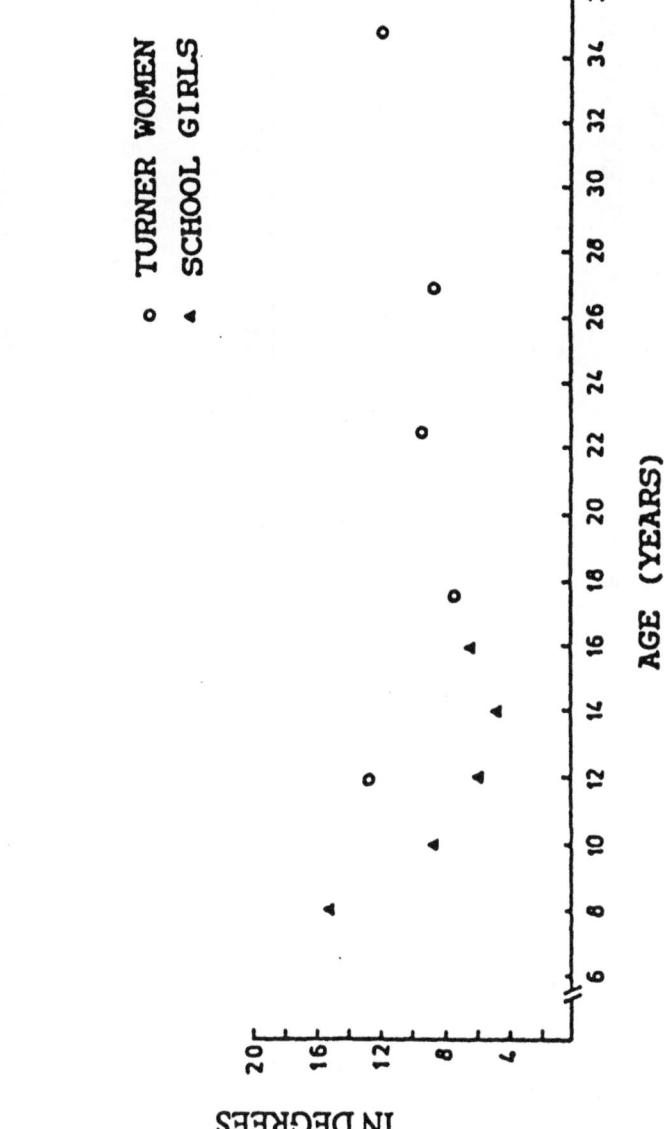

Figure 5.1 Changes with age in group mean unweighted phi value for Turner women and school girls.

106

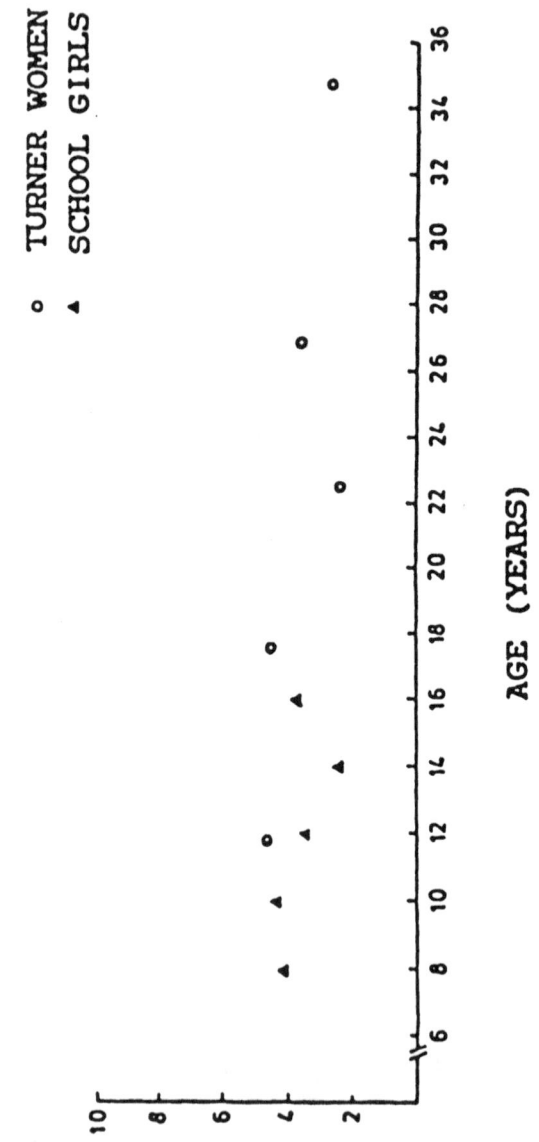

Figure 5.2 Changes with age in group mean response consistency (sigma) for Turner women and school girls.

Changes in weighted *phi* scores. Table 5.1 shows the number and percentages of subjects in each of the age groups who were unable to benefit from vestibular-somesthetic information and accordingly could not dispense with the misleading frame tilt information (i.e., who obtained a weighted *phi* value significantly different from zero, $p < .05$). The proportion of the TS women with a significant weighted *phi* score did not change much over time. More than 70% of the TS women put significant emphasis on the misleading frame tilt information throughout the age range studied. By contrast, a clear age change can be seen in the unaffected girls. Thus, while 87% of the 8-year-old girls emphasized frame tilt information, this proportion decreased in a fairly linear way with age to less than 50% at 16 years. The distribution of weighted *phi* scores for TS women differs significantly from that of the unaffected girls ($X^2 = 10.87$, $p < .05$).

Changes in type of spatio-perceptual strategy used. Table 5.2 presents changes with age in spatio-perceptual strategy used. A clear and consistent age change in response consistency can be observed in the unaffected girls, with a smaller proportion using one of the unsystematic D strategies with increasing age. The proportion of TS women applying an unsystematic D strategy also became smaller with age. However, this age change was delayed relative to that of the unaffected girls. While the proportion of unaffected girls who applied one of the unsystematic D strategies decreased to below 50% in the age range of 10 to 12 years, a similar decrease in the use of D strategies by TS women is first noticeable years later.

Another age difference in the distribution of strategies between TS women and the unaffected girls is shown in Table 5.2. Almost all (93%) of the 8-year-old unaffected girls used non-optimal strategies (i.e., either an unsystematic D strategy or a systematic but strongly optically dominated C strategy). However, at the ages of 14 and 16 less than half of the unaffected girls (47% and 40%, respectively) applied a C or a D strategy. In contrast, 71% of the 15- to 19-year-old TS women, and 75% of those over 30 were still using these non-optimal spatio-perceptual strategies. The age distribution of the TS women using non-optimal D and C strategies in preference to the optimal or near-optimal B an A strategies differs significantly from that of the unaffected girls ($X^2 = 35.76$, $p < .001$).

## Discussion

The present analyses of RFT performance yielded four observations. First, the proportion of unaffected girls that could deemphasize the potential impact of misleading frame tilt information (weighted *phi*) by effective reference to vestibular-somesthetic information becomes smaller with age. Second, the number of TS women who either responded inconsistently to the frame tilt information, or were systematically misled by it did not diminish with age. Third, the new way of analyzing performance in the RFT indicated that normal age changes in spatio-perceptual

**TABLE 5.1**
**Mean Weighted Phi Values for TS Women and Schoolgirls**

| AGE | TURNER WOMEN | | | | | SCHOOLGIRLS | | GROUP | | |
|---|---|---|---|---|---|---|---|---|---|---|
| | MEAN | SD | N | n | n* | n | n* | N | AGE | |
| < 15 YR. | $12^{11}$ | $2^5$ | 14 | 2 (14%) | 12 (86%) | 2 (13%) | 13 (87%) | 15 | 8 YR | |
| 15 - 19 YR. | $17^7$ | $1^6$ | 7 | 2 (29%) | 5 (71%) | 6 (40%) | 9 (60%) | 15 | 10 YR | |
| 20 - 24 YR. | $22^6$ | $1^4$ | 11 | 2 (18%) | 9 (82%) | 7 (47%) | 8 (53%) | 15 | 12 YR | |
| 25 - 29 YR. | $27^0$ | $1^5$ | 9 | 2 (22%) | 7 (78%) | 8 (53%) | 7 (47%) | 15 | 14 YR | |
| > 30 YR. | $34^9$ | $3^5$ | 4 | 0 (0%) | 4 (100%) | 8 (53%) | 7 (47%) | 15 | 16 YR | |

*p < .05

**TABLE 5.2**

**Changes with Age in Spatio-Perceptual Strategies Used by TS Women and Schoolgirls**

**TS WOMEN**

| AGE | MEAN | SD | N | OPTIMAL OR NEAR-OPTIMAL STRATEGY | | NON-OPTIMAL STRATEGY | |
|---|---|---|---|---|---|---|---|
| | | | | A | B | C | D |
| < 15 YR | $12.1$ | $2.5$ | 14 | 7 (50%) | 4 (29%) | 0 (0%) | 3 (21%) |
| 15 - 19 YR | $17.7$ | $1.6$ | 7 | 4 (57%) | 1 (14%) | 1 (14%) | 1 (14%) |
| 20 - 24 YR | $22.6$ | $1.4$ | 11 | 3 (27%) | 3 (27%) | 2 (18%) | 3 (27%) |
| 25 - 29 YR | $27.0$ | $1.5$ | 9 | 2 (22%) | 2 (22%) | 1 (11%) | 4 (44%) |
| > 30 YR | $34.9$ | $3.5$ | 4 | 1 (25%) | 2 (50%) | 0 (0%) | 1 (25%) |

**SCHOOLGIRLS**

| OPTIMAL OR NEAR-OPTIMAL STRATEGY | | NON-OPTIMAL STRATEGIES | | N | AGE | GROUP |
|---|---|---|---|---|---|---|
| A | B | C | D | | | |
| 0 (0%) | 1 (7%) | 5 (33%) | 9 (60%) | 15 | 8 YR | |
| 1 (7%) | 0 (0%) | 5 (33%) | 9 (60%) | 15 | 10 YR | |
| 6 (40%) | 3 (20%) | 1 (7%) | 5 (33%) | 15 | 12 YR | |
| 5 (33%) | 3 (20%) | 1 (7%) | 6 (40%) | 15 | 14 YR | |
| 4 (27%) | 2 (13%) | 4 (27%) | 5 (33%) | 15 | 16 YR | |

Stat A: Systematic vestibular dominated compromise strategy.
Stat B: Systematic vestibular compromise strategy.
Stat C: Systematic optic dominated compromise strategy.
Stat D: Unsystematic strategy.

strategies in the RFT follow this pattern: Most 8- and 10-year-old unaffected girls use either an inconsistent non-optimal D strategy or a consistent optically dominated, non-optimal C strategy. However, by age 12 many girls have already become able to use a near-optimal compromise B strategy or even to master an optimal vestibular-somesthetic dominated A strategy. Fourth, it was observed again that TS women do not normally pass through these phases. The majority of TS women persist in using non-optimal D or C strategies for a long time after a considerable proportion of unaffected girls have turned to the more successful B or A strategies.

Recall that about half of the TS women studied had received cyclic estrogen-gestagen therapy for various periods for time. Duration of treatment was shown to be related to the expression of spatial ability in such a way that short-term treated TS women had normal spatial ability, while untreated and long-term treated individuals had equally low spatial ability (Nyborg & Nielsen, 1981a). However, it is not likely that differences in duration of treatment can explain the present Turner age distribution, because the three treatment types were nearly equally represented in the various age groups. This reduces the error score equally in the age groups and leads to a conservative estimate of the size of the error score in TS women. An exception to the equal representation of differently-treated TS women is the youngest group of Turner girls in which none were treated before the age of 14. While this group contained some of the short-term treated girls, the average error score of the total group was very high.

The observations suggest that during the prepubertal period, most young children developed ways of systematically and effectively handling the information available to them in the RFT situation. Thus, although first being confused or overwhelmingly misguided systematically by the illusion creating optical tilt information, they gradually were able to make a stable compromise between conflicting optical and vestibular-somesthetic sources of information or a clear perceptual preference for one type. After puberty, 40% of the unaffected girls were able to maximally exploit the reliable vestibular-somesthetic information in the RFT situation. This enabled these individuals to cope adequately with the perceptual conflict. TS women do not seem to pass normally through these phases. They either stagnate in an early phase of spatio-perceptual strategy development or develop at a much slower pace than normal girls.

A tendency for some unaffected girls to regress in spatial ability to a prepubertal stage sometime between 14 and 16 years of age was observed in the present study as it has been in other studies (e.g., Witkin, Goodenough, & Karp, 1967). Albeit slight, this regression has interesting theoretical implications because it appears at the same time as the pubertal regression in areas believed to be related to spatial ability, such as nonverbal IQ, mathematics, physics, and science achievement (for a review, see Nyborg, 1983a). Obviously, we need much more extensive, longitudinal data before we can confirm that the slight, cross-sectionally-derived age regression in RFT performance found in this study reflects a genuine developmental regression. However, the youngest (then 8-year-old) girls from the large cross-sectional study have been retested together with two other groups in a cohort-sequential study with the RFT at regular intervals since 1976. A preliminary analysis of these longitudinal data indicates that the regression in the present study and that observed by the Witkin group do in fact reflect a genuine developmental trend.

The findings raise a number of questions. Have TS women really lost the genetic potential for the full expression of spatial ability with the loss of X chromosome material? I do not think so, and I defend this position in the following sections. However, if TS women have the genes necessary for spatial ability, it follows that the slowdown or the arrest in spatio-perceptual development has to be explained by ontogenetic factors capable of inhibiting the expression of the genes. This also raises questions. Are these ontogenetic factors the same as those responsible for pubertal regression in the spatio-perceptual development of normal girls as estimated by the RFT (Witkin et al., 1967)? I think they are. What, then, is the nature of these factors and how do they exert their influence?

In an attempt to answer these questions I examine first the results of previous research on the localization of agents believed to be important for the expression of spatial ability. I go on to propose a neuroendocrinological alternative.

## FACTORS INFLUENCING THE EXPRESSION OF SPATIAL ABILITY

There have been many diverse attempts to determine the factors that influence the expression of spatial ability (see Chapter 3 in this volume for a more thorough review). Waber (1976, 1977a, 1977b, 1979a, 1979b) speculated that cerebral reorganization at puberty inhibits spatial ability, and furthermore, she suggested that defective visuo-motor coordination and difficulties with manual sequences are related to inhibition of the expression of spatial ability. McGlone (1985) found a positive correlation between decreased somatosensory threshold of the left palm and spatial ability, but was unable to support the conclusion that the neuropsychological deficits of TS women are due to a focal brain dysfunction. Netley (1977) conjectured that the cerebral hemispheres of TS women are bilaterally organized more than usual for both verbal and spatial ability. The intrahemispheric verbal-spatial competition is believed to thwart the expression of spatial ability (Levy, 1969). Nevertheless, there is still a lack of evidence that would unequivocally link spatial ability with degree of cerebral asymmetry in TS women (e.g., McGlone, 1985; Waber, 1979b) or in unaffected individuals (Kimura, 1987).

The expression of spatial ability has also been explained by an X-linked, recessive gene theory (O'Connor, 1943; Stafford, 1961, 1963, 1972). However, more recent evidence does not support the theory (Boles, 1980; Bouchard & McGee, 1977; Corley, DeFries, Kuse, & Vandenberg, 1980; DeFries et al., 1976; Guttman, 1974; Loehlin, Sharan, & Jacoby, 1978; McGee, 1979, 1982; Nyborg & Nielsen, 1981b; Vandenberg & Kuse, 1979). In fact, the X-linked, recessive gene theory predicts that TS women should have high spatial ability; but this prediction obviously runs counter to all available evidence (e.g., Nyborg & Nielsen, 1981a).

According to Moor (1967), excess of X chromosome material suppresses higher cognitive functions in man. However, it could just as well be argued that lack of X chromosome material suppresses the expression of spatial ability in TS women. Moreover, the expression of spatial ability does not vary much within Turner syndrome across karyotypes differing with respect to the amount of X chromosome material present (Nielsen et al., 1977). Suggestions that the sheer amount of X chromosome material or heterochromatin may have impact on a phenotypical trait begs questions about the location of action and about details of the mechanisms (Nyborg, 1986b).

On the whole, the research to date still leaves open the question of whether TS women have lost the genetic potential for normal spatial ability. Furthermore, it is not particularly helpful in illustrating the nature and the mechanisms of the ontogenetic factor(s) that are powerful enough to diminish the expression of spatial ability. Finally, the research has not yet clarified whether the factors controlling the expression of spatial ability in TS women are identical to those controlling it in unaffected women. Consequently, the field is in need of both a comprehensive theory and research designed specifically to solve these intricate problems.

## SEX HORMONES AND THE EXPRESSION OF SPATIAL ABILITY

In an early attempt to deal simultaneously with these difficult questions I suggested that (1) most, if not all, sex-dimorphic traits are sex-limited rather than sex-linked, and (2) estradiol (a biologically potent sex steroid in the estrogen group) plays a major role in regulating the expression of many of them, partly by modulating DNA products on autosomal chromosomes and partly by influencing the working conditions of neurotransmitters (Nyborg, 1979, 1981, 1983a, 1983b, 1984, 1985, 1986a, 1986b, 1987a, 1987b, 1987c, 1987d, 1987e).

In the following sections I examine four lines of evidence indicating that life-history variations in estradiol do indeed cause alterations in the expression of spatial ability and of other abilities as well: (1) sex steroid changes at puberty; (2) sex hormone substitution therapy of TS women; (3) the effects of menstrual changes; and (4) sex hormone substitution therapy of postmenopausal women. The effects of these four steroid perturbations on the expression of spatial ability are examined with regard to predictions resulting from the General Trait Covariance-Androgen/Estrogen (GTC-A/E) balance model illustrated in Figure 5.3 (Nyborg, 1987b).

### Changes at Puberty

Continuity of development characterizes the expression of spatial ability as measured by the RFT during the extensive prepubertal period. Despite tremendous individual differences, most children tend to take up a stable position within the spatial ability dimension relative to the prototypic developmental trend (Witkin et al., 1967). This picture of smooth development of spatial ability changes abruptly at puberty, when discontinuity of development becomes a prominent feature (Witkin et al., 1967). The onset of discontinuity coincides with radical changes in plasma estradiol concentrations. The pubertal discontinuity typically takes one of two forms. Either the developmental curve reaches an asymptote, in which case the child's relative position within the ability dimension remains constant, or an actual reduction of ability takes place. In the latter case each child changes its relative position in accordance with the force of the inhibiting agent.

Consistent with the GTC-A/E balance model the pubertal modulation of spatial ability is moderately related to the degree of development of secondary sexual characteristics. Thus, while late maturing children have been observed to reach their spatial asymptote late in puberty, early maturing children seem characterized by a tendency toward postpubertal inhibition of the expression of spatial ability. The pubertal inhibition or, more appropriately, regression of spatial ability in early maturing children seems to take place irrespective of sex. This regression can also explain why highly androgenized men and highly estrogenized women tend to exhibit low spatial ability, while late maturing "androgynous" men and women tend to show high spatial ability (Crockett & Petersen, 1985; Linn & Petersen, 1985; Maccoby & Jacklin, 1974; Petersen, 1976, 1979; Waber, 1976, 1977a, 1977b, 1979a).

According to the model, the sexually most differentiated individuals "overshoot" the range of brain estradiol concentrations for the optimal expression of spatial ability, while the sexually less differentiated individuals remain below or within this range. In reality, things are more complicated. For example, we know that both men and women are able to convert (aromatize) testosterone to estradiol, and that they probably do so in different amounts in different tissues at different ages. However, I shall minimize the account of biochemical details in order to illustrate better the major principles of the GTC-A/E balance model. (In addition to dealing with covariant development of hormonal, bodily, and spatial ability traits, the GTC-A/E model also strives to explain covariant personality development [Nyborg, 1983a, 1983b, 1984, 1987a, 1987b,

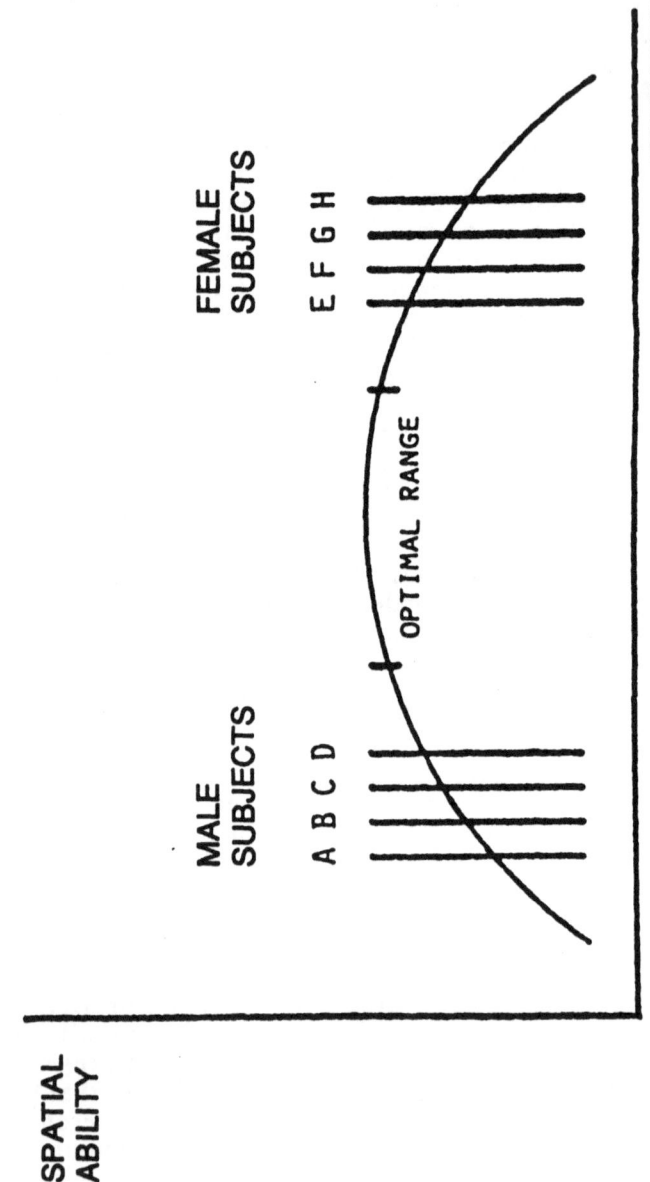

**Figure 5.3** Expression of spatial ability at different plasma estradiol concentrations as predicted by the General Trait Covariance-Androgen/Estrogen (GTC-A/E) balance model.

1987c, 1987d, 1987e], although an account of this aspect of the model is outside the scope of this chapter.)

### Sex Hormone Substitution Therapy in Turner Women

Another perturbating life-history event is that of receiving supplementary sex hormone therapy. Effects of estrogen therapy can be studied in a quasi-experimental design with TS women who receive treatment to compensate for their unusually low secretion of gonadal hormones. We performed a retrospective study of such women, in which the duration of treatment figured as the independent variable (Nyborg & Nielsen, 1981a). We found that TS women receiving short-term cyclic estrogen/gestagen treatment for between three months and two years (1.1 year average) obtained spatial ability scores identical to those of their age-matched chromosomally normal sisters on a number of spatial ability tasks, including the RFT. The groups of untreated TS women and long-term treated TS women (with an average of about eight years of treatment) both scored way below the short-term treated women.

We recognized that the methodological weaknesses inherent in such a retrospective study seriously limit the findings. Nevertheless, the observations make sense under the assumptions that (1) untreated TS women do not secrete enough gonadal hormones to reach the estradiol concentrations necessary for proper brain development and for the optimal expression of spatial ability, and (2) short-term cyclic estrogen-gestagen therapy exerts beneficial organizational effects on somatic as well as on estrophilic brain tissues, albeit in a highly atypical manner. In addition, the treatment brought the estradiol concentrations close to or within the range for optimal expression and thus exerted a positive activational effect on the brain for the expression of spatial ability. Given that high plasma estradiol concentrations impede the expression of spatial ability, and that high doses (Toran-Allerand, 1984) and perhaps also prolonged treatment can have neurotoxic effects, the GTC-A/E model suggests that the long-term treated TS women either suffered under steroid brain intoxication or overshot the optimal range, and thus suffered an estradiol-induced inhibition of the expression of spatial ability. The doses used for the long-term cases were likely to have been excessive by today's standards. However, whether the actual treatment had neurotoxic effects remains to be confirmed. Fortunately, there is now an acute awareness of the need to drastically reduce the amount of estradiol prescribed in order to avoid unwanted side effects.

With these caveats in mind, I think the minimalist conclusion is that the above-mentioned findings are at least compatible with the idea that spatial ability is a sex-limited trait, the expression of which can be regulated up or down as a function of estradiol uptake. Another permissible conclusion is that TS women actually possess the genes required for the full expression of spatial ability irrespective of their well-documented lack of X-chromosome material. This conclusion rests upon the observations that short-term treated TS women obtain a spatial ability score identical to that of their sisters with a normal X-chromosome complement, and that short-term treated TS women with a 45,X karyotype obtain spatial ability scores identical to those of short-term treated TS women with other karyotypes.

## Perturbations Associated with Menstruation

Menstruation is hormonally a profoundly perturbating monthly event. For example, the ovarian secretion of estradiol changes markedly from a no-production state at menstruation to a peak value at ovulation. This is a typical situation in which the GTC-A/E balance model allows for a number of testable predictions about the expression of spatial ability. For example, it can be determined whether the expression of spatial ability is maximal at the time when the plasma estradiol concentration is minimal, and vice versa (see Figure 5.3). In other words, the GTC-A/E balance model predicts that the ordinary expression of female spatial ability will cycle dynamically up and down as an inversely related function of the monthly downs and ups in the availability to the brain of estradiol. This cycling phenomenon has been observed in several studies where it was found that spatial ability was high when estrogen was low, and vice versa (Anderson, 1972; Dor-Shav, 1976; Ho, Gilger & Brink, 1986; Klaiber, Broverman, Vogel, & Kobayashi, 1974). Recently, Hampson and Kimura (1987) confirmed this cycling nature of the expression of spatial ability in an RFT and embedded-figures test study of normally menstruating women.

It has been suspected for some time that fine motoric abilities, some of which women excel in compared to men, may also be under hormonal influence. For example, using a double-blind crossover, Latin-square tap test design, Itil and Herrmann (1978) demonstrated that estrogen influences motor skills by finding that their estrogen treatment group achieved significantly more taps than a placebo group in 32-second test periods at both one and three hours after medication. Moreover, Hampson and Kimura (1987) were able to show that some manual abilities cycle with variations in estradiol in such a way that when manual skills were at their highest, spatial skills were at their lowest, and vice versa.

In fact, the GTC-A/E model presupposes such a dynamic dissociation between the cyclic changes in the expression of spatial and manual skills, as the basic assumption of the GTC-A/E model is that all traits showing sexual dimorphism can be ascribed primarily to permanent or to transient effects of gonadal hormones (Nyborg, 1981, 1984). From this basic rule it follows that traits whose expression are influenced by transient activational hormonal effects will change whenever the hormones change, while traits dependent on organizational hormonal effects will be present phenotypically in a more stable manner. However, it should not be forgotten that the expression of some traits may depend on a combination of organizational and activational effects, while the expression of other traits may seem organizationally stabilized only because the level of activational gonadal hormones remains constant. For example, women on oral contraceptives show less of the phase-related changes in the expression of a number of traits that presumably depend on activational effects (e.g., Bardwick, 1976; Diamond, Diamond, & Master, 1972; Wuttke et al., 1975).

### Perturbations at Menopause

A glance at the curvilinear GTC-A/E balance model (Figure 5.3) will immediately raise the expectation that postmenopausal women who have experienced a decline in plasma estradiol concentrations will obtain a higher spatial ability score than they did in their premenopausal-postpubertal life-span. This prediction has yet to be tested. However, a number of climacterical women suffer from various postmenopausal symptoms that can be eased by estrogen replacement therapy. Rosenthal and Kimura (1987) exploited this situation by examining the effects of gonadal hormones on the expression of spatial and motor abilities. They localized a group of postmenopausal women who were on an on-off estrogen replacement therapy and found that while the expression of spatial ability (figure disembedding task) and praxic motor ability (manual-sequence task) were high in the estrogen treatment phase, the converse occurred in the off-therapy phases when steroid levels were believed to be lower.

When considered together, these four lines of evidence strongly suggest that certain spatial and fine motoric abilities are sex-limited and that estradiol regulates their expression. Moreover, the GTC-A/E balance model seems remarkably efficient in accounting for the intricate dynamic and static aspects of the quantitative relationships among estradiol, body maturation, and spatial and motor abilities. That is, in terms of relatively simple principles, the model can explain (1) the expression of low spatial ability encountered in untreated TS women; (2) the expression of a normal level of female spatial ability in short-term treated TS women; (3) the expression of low spatial ability in long-term treated TS; (4) the regression in spatial ability in unaffected girls at puberty; (5) the debut of the often observed adult sex-related difference in spatial ability; (6) the dynamic menstrual and postmenopausal changes in the expression of a number of abilities in adult women, along with the oral contraceptive or noncyclic fixation of these abilities; and (7) the noncyclic nature of spatial and motoric abilities in men.

I have suggested elsewhere (Nyborg, 1981) that there exists a range of estradiol values optimal for the expression of certain verbal abilities (e.g., verbal fluency), and that this range is inversely related to that for the expression of spatial ability. This suggestion fits nicely with Lynn's (1987) recent argument for the existence of a negative correlation between spatial and verbal abilities once "g" is controlled. Lynn speculated that the correlation somehow relates to maturational rate.

In their study of postmenopausal women, Rosenthal and Kimura (1987) actually tested the prediction of the GTC-A/E balance model that verbal memory and verbal fluency scores were affected by higher levels of estrogen in the on-therapy phase as compared to the off-therapy phase; however, Hampson (1986) found the same phase-related patterns of change for speeded manual coordination in her speech-articulatory data. Perhaps the substitution-therapy-related level of plasma estrogen concentration in the on-therapy phase of the Rosenthal and Kimura study was too low to modulate the expression of verbal ability.

The assumption of the GTC-A/E balance model that genetic potential for the expression of spatial ability is coded in DNA material located on autosomal chromosomes also appears to be supported by, or at least not incompatible with, available data on TS subjects. Neither the complete absence of one X chromosome in all cell lines, nor abnormalities of the second X chromosome interferes with the full expression of spatial ability in TS women, so long as their estradiol deficiency is corrected (Nyborg & Nielsen, 1981a).

## THE NEUROPHYSIOLOGY OF SPATIO-PERCEPTUAL STRATEGIES

Perception of the upright in the RFT depends on whether the optical information can be successfully integrated with the gravitational field information. Optical information alone is insufficient, because the retinal information about directions in space cannot be evaluated unless the position of the eye in the environmental coordinate system is known. The undisturbed access to gravitational information enables the subject to collect such information and to establish and maintain a veridical perception of the upright. Under normal terrestrial conditions this poses no problem because the equivalent optical and gravitational information is congruent. However, in the RFT the optical and the gravitational information is experimentally rendered incongruent by tilting the visual framework. Therefore, the subjects must solve the perceptual conflict between the optical and the vestibular-somesthetic information before successful perception of the upright can be accomplished. Solution requires effective cross-modal matching of information arising from different perceptual systems, and the spatio-perceptual strategies are believed to reflect the kind of solution to this conflict found in the RFT.

Obviously, the complex optic-vestibular-somesthetic, cross-modal matching of afferent information required in the RFT situation must be accomplished by the central nervous system (Bischof, 1974a; Nyborg, 1977). The cross-modal matching causes problems for TS women who suffer from an estradiol deficit, but not for those having received short-term supplementary estrogen therapy. This means, logically, that untreated TS women either have less access to or are less able to process the bodily information derived from the gravitational field force and collected through the vestibular and somesthetic systems.

Available neurophysiological and neurological evidence does not allow us to draw final conclusions about the localization of the putative brain disturbances responsible for the abnormal spatio-perceptual development in TS women (e.g., McGlone, 1985; Reske-Nielsen, Christensen, & Nielsen, 1982). However, whatever the locus of the brain malfunctioning, a case can be made for testing the hypothesis that the damage is a more or less direct result of their abnormally low, plasma estradiol concentration.

We still have a long way to go before we know what goes wrong with brain development in untreated TS women. Nevertheless, we have reasons to remain optimistic. Provided that the assumptions of the GTC-A/E balance model are valid (i.e., that DNA material required for the build-up of brain tissues subserving spatial information processing is located on autosomal chromosomes, and further that the expression of spatial ability is regulated by estradiol), we should be able to correct the spatio-perceptual deficits in TS women by stimulating brain growth via the same estrogen substitution therapy that is called for to stimulate the development of their secondary sexual bodily characteristics. To grasp the details of this possibility, brain development and functioning must be closely monitored before, during, and after estrogen treatment.

This goal would be easier to attain if we succeed in establishing a protocol for an internationally coordinated, large-scale study to examine the structural and neurophysiological changes that take place in estrophilic brain areas as a function of low-dosage estrogen substitution therapy in TS women. Such a protocol should involve modern regional cerebral blood-flow techniques, as well as NMR brain imaging and spectroscopical studies, in addition to perceptual, intellectual, and personality test batteries. Inclusion also of anthropometric measures would add information about how gonadal hormones manage to procure covariant body, brain, and behavioral-trait development. Finally, the protocol should pay due respect to individual differences in preceding prenatal and postnatal endogenous and exogenous endocrine conditions, and should acknowledge information about patterns of inheritance for the traits in question.

## NOTES

1. When Asch and Witkin (1948) constructed the RFT, they assumed that the test measured individual differences in how well people solve the experimentally-induced conflict between the visual and the postural information regarding the physical vertical. Later,

the Witkin group (e.g. Witkin et al., 1962/1974) became conceptually more embracing and constructed an elaborated psychological framework to account for performance in the RFT. They postulated that successful performance in the RFT directly reflects the degree of field independence in the sense of being capable of freeing oneself from contextual (read visual) perceptual influences and being able to "disembed" the rod when adjusting it to a physically vertical position. When it gradually became evident that field-independent persons also tend to behave in an independent manner in social settings, and that field-dependent persons tend to be more easily influenced by their social surroundings, much like they are by the perceptual context, the Witkin group invoked the higher order construct of psychological differentiation. It became customary to think that the simple error scores in the RFT more or less directly reflect the extent of psychological differentiation of personality. The four variables comprising the content analysis are believed to reflect the functioning of neurophysiological factors underlying spatial ability rather than personality characteristics.

2. It is interesting to note that the Witkin group (Witkin & Goodenough, 1981) also came to a similar conclusion that, ironically, is very similar to the original Asch and Witkin (1948) interpretation of RFT performance.

## REFERENCES

Anderson, E. I. (1972). Cognitive performance and mood change as they relate to menstrual cycle and estrogen level. Dissertation Abstracts, 33, 1758-B.

Arbuthnot, J. (1972). Cautionary note on measurement of field independence. Perceptual and Motor Skills, 35, 479-488.

Asch, S. E. & Witkin, H. A. (1948). Studies in space orientation. I. Perception of the upright with displaced visual fields. Journal of Experimental Psychology, 38, 325-337.

Bardwick, J. M. (1976). Psychological correlates of the menstrual cycle and oral contraceptive medication. In: E. J. Sachar (Ed.), Hormones, behavior, and psychopathology. Raven Press: New York.

Bischof, N. (1974). Optic-Vestibular orientation to the vertical. I: H. H. Kornhuber (Ed.). Vestibular systems, Part 2: Psychophysics, applied aspects and general interpretations, (pp. 155-190). Berlin: Springer-Verlag.

Boles, D. B. (1980). X-Linkage of spatial ability: A critical review. Child Development, 51, 625-635.

Bouchard, T. J., Jr., & McGee, M. G. (1977). Sex differences in human spatial ability: Not an X-linked recessive gene effect. Social Biology, 24, 332-335.

Corley, R. P., DeFries, J. C., Kuse, A. R., & Vandenberg, S. G. (1980). Familial resemblance for the identical blocks test of spatial ability: No evidence for X linkage. Behavior Genetics, 10, 211-216.

Crockett, L. J., & Petersen, A. C. (1985). Pubertal status and psychosocial development: Findings from the early adolescence study. In R. M. Lerner & T. T. Foch (Eds.), Biological-psychosocial interactions in early adolescence: A life-span perspective. Hillsdale, NJ: Erlbaum.

DeFries, J. C., Ashton, G. C., Johnson, R. C., Kuse, A. R., McClearn, G. E., Mi, M. P., Rashad, M. N., Vandenberg, S. G., & Wilson, J. (1976). Parent-offspring resemblance of specific cognitive abilities in two ethnic groups. Nature, 261, 131-133.

Diamond, M., Diamond, A. L., & Master, M. (1972). Visual sensitivity and sexual arousal levels during the menstrual cycle. The Journal of Nervous and Mental Disease, 155, 170-176.

Dor-Shav, N. K. (1976). In search of pre-menstrual tension: Note on sex-differences in psychological-differentiation as a function of cyclical physiological changes. Perceptual and Motor Skills, 40, 683-693.

Gruen, A. (1957). A critique and re-evaluation of Witkin's perception and perception-personality work. The Journal of General Psychology, 56, 73-93.

Guttman, R. (1974). Genetic analysis of analytical spatial ability: Raven's Progressive Matrices. Behavior Genetics, 4, 273-284.

Haller, O. (1981). A new procedure for determining components of field dependency. Perceptual and Motor Skills, 53, 795-798.

Haller, O., & Edgington, E. S. (1982a). Scoring rod-and-frame tests: Quantitative and qualitative considerations. Perceptual and Motor Skills, 55, 587-593.

Haller, O., & Edgington, E. S. (1982b). Interpretations of rod-and-frame test scores: An application of pattern analysis. Perceptual and Motor Skills, 54, 1339-1342.

Hampson, E. (1986). Variations in perceptual and motor performance related to phase of the menstrual cycle. Canadian Psychology, 27, 268.

Hampson, E., & Kimura, D. (June, 1987). Reciprocal effects of hormonal fluctuations on human motor and perceptuo-spatial skills. Research Bulletin, (No. 656). Department of Psychology, The University of W. Ontario, London, Canada, 1-20.

Ho, H. Z., Gilger, J. W., & Brink, T. M. (1986). Effects of menstrual cycle on spatial information-processes. Perceptual and Motor Skills, 63, 743-751.

Itil, T. M., & Herrmann, W. M. (1978). Effects of hormones on computer-analyzed human electroencephalogram. In M. A. Lipton, A. Diamscio, & K F. Killam (Eds.), Psychopharmacology: A generation of progress. New York: Raven Press.

Jahoda, G., & Nielsen, I. (1986). Nyborg's analytical Rod-and-Frame scoring system: A comparative study in Zimbabwe. International Journal of Psychology, 21, 19-29.

Kimura, D. (1987). Are men's and women's brains really different? Canadian Psychology, 28, 133-147.

Klaiber, E. L., Broverman, D. M., Vogel, W., & Kobayashi, Y. (1974). Rhythms in Plasma MAO Activity, EEG, and Behavior during the Menstrual Cycle. In M. Ferin, F. Halberg, R. M. Richart, & R. L. van de Wiele (Eds.), Biorhythms and human reproduction, (pp. 353-367). New York: John Wiley.

Lester, G. (1968). The Rod-and-Frame Test: Some comments on methodology. Perceptual and Motor Skills, 26, 1307-1314.

Lester, G. (1971). Subjects' assumptions and scores on the Rod-and-Frame test. Perceptual and Motor Skills, 32, 205-206.

Levy, J. (1969). Possible basis for the evolution of lateral specialization of the human brain. Nature, 224, 614-615.

Linn, M. C., & Petersen, A. C. (1985). Emergence and characterization of sex differences in spatial ability: A meta-analysis. Child Development, 56, 1479-1498.

Loehlin, S., Sharan, S., & Jacoby, R. (1978). In pursuit of the "spatial gene": A family study. Behavior Genetics, 8, 27-42.

Lynn, R. (1987). The intelligence of the Mongoloids: A psychometric, evolutionary, and neurological theory. Personality and Individual Differences, 8, 813-844.

Maccoby, E. E., & Jacklin, C. N. (1974). The psychology of sex differences. Stanford, CA: Stanford University Press.

McGarvey, B., Maruyama, G., & Miller, N. (1977). Scoring field dependence: A methodological analysis of five Rod-and-Frame scoring systems. Applied Psychological Measurement, 1, 433-446.

McGee, M. G. (1979). Human spatial abilities: Psychometric studies and environmental, genetic, hormonal, and neurological influences. Psychological Bulletin, 86, 889-917.

McGee, M. G. (1982). Spatial abilities: The influence of genetic factors. In M. Potegal (Ed.), Spatial abilities: Development and physiological foundations, (pp. 199-222). New York: Academic Press.

McGlone, J. (1985). Can spatial deficits in Turner's syndrome be explained by focal CNS dysfunction or atypical speech lateralization. Journal of Clinical and Experimental Neuropsychology, 7(4), 375-394.

Moor, L. (1967). Niveau intellectuel et polygonosomie: Confrontation du caryotype et du niveau mental de 374 malades dont le caryotype comporte un exces de chromosomes X ou Y. Revue de Neuropsychiatrie infantile, 15, 325-348.

Netley, C. (1977). Dichotic listening of callosal agensis and Turner's syndrome patients. In S. J. Segalowitz & F. A. Gruber (Eds.), Perspectives in neurolinguistics and psycholinguistics language development and neurological theory. New York: Academic Press.

Nielsen, J., Nyborg, H., & Dahl, H. (1977). Turner's syndrome. A psychiatric-psychological study of 45 women with Turner's syndrome, compared with their sisters and women with normal karyotype, growth retardation, and primary amenorrhoea. Aarhus: Acta Jutlandica, Medicine Series 21, Aarhus.

Nyborg, H. (1971a). Tactile stimulation and perception of the vertical: I. Effects of diffuse versus specific tactile stimulation. Scandinavian Journal of Psychology, 12, 1-3.

Nyborg, H. (1971b). Tactile stimulation and perception of the vertical: II. Effects of field dependency, arousal, and cue function. Scandinavian Journal of Psychology, 12, 135-143.

Nyborg, H. (1972). Light intensity and perception of the vertical: Two experiments with the rod-and-frame test. Scandinavian Journal of Psychology, 13, 1-13.

Nyborg, H. (1974a). A method for analysing performance in the rod-and-frame test. I. Scandinavian Journal of Psychology, 15, 119-123.

Nyborg, H. (1974b). Light intensity in the rod-and-frame test reconsidered. Scandinavian Journal of Psychology, 15, 236-237.

Nyborg, H. (1977). The rod-and-frame test and the field dependence dimension: Some methodological, conceptual, and developmental considerations. Copenhagen: Dansk Psykologisk Forlag.

Nyborg, H. (1979). Sex chromosome abnormalities and cognitive performance. V: Female sex hormone and discontinuous cognitive development. Paper presented in the symposium on "Cognitive Studies" at the Fifth Biennial Meeting of the International Society for the Study of Behavioral Development, Lund, Sweden, June 25-29.

Nyborg, H. (1981). Hormonal correlates of spatial ability development. Paper presented at the Sixth Congress of The International Society for the Study of Behavioral Development, Toronto, Canada, August. (Symposium chaired by H. Nyborg and L. Harris).

Nyborg H. (1983a). Spatial ability in men and women: Review and new theory. Advances in human research and therapy, Vol. 5, (pp. 39-140). Monograph Series, London: Pergamon Press.

Nyborg, H. (1983b). Covariant intellectual and personality development in 14 hormonally different groups: A psychoneuroendocrinological model. Paper presented at the "Inaugural Meeting of the International Society for the Study of Individual Differences." London, July.

Nyborg, H. (1984). Performance and intelligence in hormonally-different groups. In: G. J. de Vries, J. P. C. de Bruin, H. B. M. Uylings, & M. A. Corner (Eds.). Sex differences in the brain. Progress in brain research. Vol. 61, (pp. 491-508). Amsterdam: Elsevier Biomedical Press.

Nyborg, H. (1985). Orchestration of body, brain, and behavioral development. Lecture presented at the University of Calgary, Department of Psychology, Calgary, Canada, May.

Nyborg, H. (1986a). <u>Sexual differentiation of the brain</u>. Paper presented at the International Conference on "Knowledge and Learning -- Ideas in Cerebral Palsy" organized by the International Cerebral Palsy Society with Spastics Society, Athens, Greece, April.

Nyborg, H. (1986b). <u>Sex chromosomes, sex hormones, and developmental disturbances: In search of a model</u>. Paper presented at the 152nd Annual National Meeting of the American Association for the Advancement of Science, Philadelphia, USA, May 25-30.

Nyborg, H. (1987a). Individual differences or different individuals? That is the question. <u>Behavioral and Brain Sciences, 10</u>, 34-35.

Nyborg, H. (1987b). <u>Covariant trait development across races and within individuals: Differential K theory, genes, and hormones</u>. Paper presented in the symposium on "Biology-Genetics" at the Third Meeting of the International Society for the Study of Individual Differences," Toronto, Canada, June 18-22.

Nyborg, H. (1987c). Mathematics, animosity, and sex hormones. <u>Behavioral and Brain Sciences</u>. (Submitted).

Nyborg, H. (1987d). Principles of sex hormonal regulation of body, brain, and behavioral development. <u>Behavioral and Brain Sciences</u>. (Submitted).

Nyborg, H. (1987e). <u>Sex hormones, behavioral development, and reproduction rates: A covariant pattern</u>. Paper presented at the First International Capri Conference on Brain and Female Reproductive Function: Basic and Clinical Aspects, Capri, Italy, May 25-29.

Nyborg, H., & Isaksen, B. (1974). A method for analyzing performance in the rod-and-frame test. II. Test of the statistical model. <u>Scandinavian Journal of Psychology, 15</u>, 124-126.

Nyborg, H., & Nielsen, J. (1981a). Sex hormone treatment and spatial ability in women with Turner's syndrome. In: W. Schmid & J. Nielsen (Eds.), <u>Human behavior and genetics</u>, (pp. 167-182). Amsterdam: Elsevier/ North-Holland.

Nyborg, H., & Nielsen, J. (1981b). Spatial ability of men with Karyotype 47,XXY, 47,XYY, or normal controls. In: W. Schmid & J. Nielsen (Eds.), <u>Human behavior and genetics</u>, (pp. 167-182). Amsterdam: Elsevier/ North-Holland.

O'Connor, K. P., & Shaw, J. C. (1978). Field dependence, laterality and the EEG. Biological Psychology, 6, 93-109.

Petersen, A. C. (1976). Physical androgyny and cognitive functioning in adolescence. Developmental Psychology, 12(6), 524-533.

Petersen, A. C. (1979). Hormones and cognitive functioning in normal development. In M. A. Wittig & A. C. Petersen (Eds.), Sex-related differences in cognitive functioning: Developmental issues. New York: Academic Press.

Reske-Nielsen, E., Christensen, A. L., & Nielsen, J. (1982). A neuropathological and neuropsychological study of Turner's syndrome. Cortex, 18, 181-190.

Rosenthal, K., & Kimura, D. (1987). Hormonal influences on cognitive ability patterns. Research Bulletin, No. 653, March. Department of Psychology, The University of W. Ontario, London, Canada, 1-20.

Stafford, R. E. (1961). Sex differences in spatial visualization as evidence of sex-linked inheritance. Perceptual and Motor Skills, 13, 428.

Stafford, R. E. (1963). An investigation of similarities in parent-child test scores for evidence of hereditary components. Educational Testing Service, Princeton.

Stafford, R. E. (1972). Hereditary and environmental components of quantitative reasoning. Review of Educational Research, 42, 183-201.

Toran-Allerand, C. D. (1984). On the genesis of sexual differentiation of the central nervous system: Morphogenetic consequences of steroidal exposure and possible role of alpha-fetoprotein. In G. J. de Vries, J. P. C. de Bruin, H. B. M. Uylings & M. A. Corner (Eds.), Sex differences in the brain. Progress in brain research, Vol. 61. Amsterdam: Elsevier Biomedical Press.

Turner, H. (1938). A syndrome of infantilism, congenital webbed neck, and cubitus valgus. Endocrinology, 23, 566-574.

Vandenberg, S. G., & Kuse, A. R. (1979). Spatial ability: A critical review of the sex-linked major gene hypothesis. In M. A. Wittig & A. C. Petersen (Eds.), Sex-related differences in cognitive functioning: Developmental issues, (pp. 67-95). New York: Academic Press.

Waber, D. P. (1976). Sex differences in cognition: A function of maturation rate. Science, 192, 572-574.

Waber, D. P. (1977a). Sex differences in mental abilities, hemispheric lateralization, and rate of physical growth at adolescence. Developmental Psychology, 13, 29-38.

Waber, D. P. (1977b). Biological substrates of field dependence: Implications of the sex difference. Psychological Bulletin, 84, 1076-1087.

Waber, D. P. (1979a). Cognitive abilities and sex-related variations in the maturation of cerebral cortical functions. In M. A. Wittig & A. C. Petersen (Eds.), Sex-related differences in cognitive functioning, (pp. 161-186). New York: Academic Press.

Waber, D. P. (1979b). Neuropsychological aspects of Turner's syndrome. Developmental Medicine and Child Neurology, 21, 58-70.

Witkin, H. A. (1964). Origins of cognitive style. In C. Scheerer (Eds.), Cognition: Theory, research, promise. New York: Harper and Row, 172-205.

Witkin, H. A., & Asch, S. E. (1948). Studies in space orientation. IV. further experiments on perception of the upright with displaced visual fields. Journal of Experimental Psychology, 38, 762-782.

Witkin, H. A., Dyk, R. B., Faterson, H. F., Goodenough, D. R., & Karp, S. A. (1974). Psychological differentiation. Potomac, Md: Erlbaum (Originally published, 1962).

Witkin, H. A., & Goodenough, D. R. (1977a). Field dependence and interpersonal behavior. Psychological Bulletin, 84, 661-689.

Witkin, H. A., & Goodenough, D. R. (1977b). Field dependence revisited. Research Bulletin, RB-77-16.

Witkin, H. A., & Goodenough, D. R. (1981). Cognitive styles: Essence and origins. New York: International Universities Press.

Witkin, H. A., Goodenough, D. R., & Karp, S. A. (1967). Stability of cognitive style from childhood to young adulthood. Journal of Personality and Social Psychology, 7(3), 291-300.

Witkin, H. A., Lewis, H. B., Hertzman, M., Machover, K., Meissner, P., & Wapner, S. (1954). Personality through perception. New York: Harper.

Witkin, H. A., & Oltman, P. K. (1967). Cognitive Style. <u>International Journal of Neurology</u>, <u>6</u>, 119-137.

Wuttke, W., Arnold, P., Becker, D., Creutzfeldt, O., Langestein, S., & Tirsch, W. (1975). Circulating hormones, EEG, and performance in psychological tests of women with and without oral contraceptives. <u>Psychoneuroendocrinology</u>, <u>1</u>, 141-151.

# 6    Behaviour and Extra X Aneuploid States

The material presented in this chapter was collected in an ongoing multidisciplinary investigation of development in individuals with neonatally ascertained sex chromosome abnormalities (SCA). Although our SCA sample contained a number of different X and Y aneuploid states (Bell & Corey, 1974), the majority were 47,XXY males with the result that the principal focus of our investigation has been on extra X males.

In its initial stages the research was frankly empirical and simply directed toward the issue of whether behavioural anomalies could be identified in males with a supernumerary X chromosome (Stewart et al., 1979). However, as our studies progressed, our data and those of others persuaded us to develop theoretical explanations for the range of available observations. They also persuaded us to consider the possibility that processes involved in the genesis of behavioural deviance in 47,XXY males may also be operative in extra X aneuploid females.

## BASIC OBSERVATIONS

We have shown in several investigations conducted at different times that pre-pubertal and pubertal 47,XXY males have specific deficits in verbal intelligence that stand in contrast to their essentially normal, nonverbal levels of ability (Netley, 1986; Netley & Rovet, 1982a; Netley & Rovet, 1984; Stewart, Bailey, Netley, Rovet, & Park, 1986). Similar observations have been reported by others (Ratcliffe et al., 1982; Walzer, Graham, Bashir, & Silbert, 1982), although in some cases the data have suggested that the severity of the verbal deficits may lessen as cases grow older (Robinson et al., 1986). In view of the association between verbal ability and academic performance, it is not surprising that extra X males have been found to exhibit many educational difficulties and frequently have been regarded as learning disabled (Pennington, Bender, Puck, & Robinson, 1982).

Several studies have reported a second kind of behavioural anomaly in 47,XXY males. Despite difficulties in demonstrating this due to the absence of generally accepted methods of measurement, there is now a significant body of data indicating that extra X males frequently present signs of atypical interpersonal or temperamental development. For example, Walzer and his colleagues (Walzer et al., 1978; Walzer et al., 1986) reported that 47,XXY children between 6 months and 7 years were more than usually inactive, low in energy intensity, pliant, withdrawing, and unassertive. Bancroft, Axworthy and Ratcliffe (1982) reported similar findings with adolescent 47,XXY boys in that, compared to controls, they were unusually timid and unassertive.

Our data on this issue were obtained using a variety of techniques. Initially, we depended on the somewhat informal method of semistructured interviews with parents (Stewart et al., 1979). These indicated that parents typically regarded their 47,XXY sons as distinctly passive, quiet, and introverted during their pre-adolescent years. Subsequently, we have used two parent questionnaires to assess social/emotional development in our subjects: the Personality Inventory for Children (PIC) (Wirt, Seat, & Broen, 1977) in the pre-adolescent years and the Child Behavior Checklist (CBCL) (Achenbach & Edelbrock, 1981) during early adolescence. Although the PIC results did not reveal any major abnormality, they did suggest that the extra X boys were less active than chromosomally normal boys (Stewart et al., 1986). The CBCL findings provided evidence for continuing tendencies toward inactivity but also indicated that the boys were presenting signs of withdrawal, uncommunicativeness, and a number of other behavioural difficulties.

Taken together, the evidence to date suggests that 47,XXY males have significant anomalies in social/emotional development beginning early in life. Prior to puberty they typically are regarded as passive, withdrawn, and low in activity level. Although the data are less extensive, they do suggest that during puberty extra X males show more signs of maladjustment, but continue to be regarded as withdrawn and uncommunicative with low levels of social interaction.

## NEUROPSYCHOLOGICAL DATA AND THEORY

Why do 47,XXY males have specific deficits in language development and tendencies towards passivity, inactivity, and withdrawal? Our research has led us to believe that these are not unrelated issues and may depend on the common mechanism of disturbances in hemispheric organization. The evidence for this proposition comes from a series of studies which have all involved the administration of laboratory-based measures of hemispheric organization. These measures consisted of lateralized presentations of verbal and nonverbal stimuli in the auditory, tactile, and visual modalities (i.e., dichotic, dichhaptic, and half-field tachistoscopic procedures). Subject performance has been evaluated in terms of the degree to which expected right-sided advantages for verbal material and left-sided advantages for nonverbal material were observed.

The results obtained in a study comparing the asymmetries in performance of pre-pubertal 47,XXY boys with age-matched controls indicated that the extra X boys had diminished left hemisphere specialization for language and enhanced right hemisphere specialization for nonverbal processing (Netley & Rovet, 1984). Since then we have obtained data that demonstrate that the anomalies of hemispheric organization of 47,XXY boys prior to adolescence are related to their intellectual characteristics (Netley & Rovet, 1987). The nature of the relationship indicates that extra X males with particularly pronounced right hemisphere specialization for nonverbal processing have the most severe language deficits.

Our investigations have also suggested that hemispheric-based processes are relevant to the atypical social/emotional characteristics of 47,XXY males. We have examined this issue at two stages in the development of our subjects. Using multiple regression procedures, our methodology has consisted of assessing the degree to which individual differences in social/emotional functioning are associated with hormone levels (testosterone, estradiol, follicle-stimulating hormone, and luteinizing hormone), intelligence (Verbal IQ and Performance IQ), composite hemispheric specialization indices (left hemisphere [LH], right hemisphere [RH], handedness) and quality of parenting. In the first analysis, we examined the relations between the variables just cited and those provided by parent responses to the PIC prior to puberty. The statistically significant results are summarized in Table 6.1 and provide evidence for two general conclusions: IQ is the first significant correlate of measures reflecting problems in achievement, development, and intellectual functioning; the integrity of left hemisphere functioning is the primary predictor of low activity and withdrawal, temperamental qualities that distinguish 47,XXY boys from others most clearly during their pre-pubertal years.

The second analysis, performed at a time when our subjects were in various stages of pubertal development, yielded somewhat different results. The significant associations are summarized in Table 6.2 and indicate that while left hemisphere functioning remains significantly related to uncommunicativeness, other factors emerge as major correlates of this trait as well as other traits. In particular, quality of parenting is strongly negatively correlated with several scales of maladjustment as are level of testosterone and degree of right hemisphere specialization.

These data indicate that the mechanisms underlying individual differences in social/emotional functioning in 47,XXY males change in the transition from pre-puberty to the pubertal years. Prior to the onset of puberty, the integrity of left hemispheric functioning is the best single correlate of those traits that distinguish extra X males from chromosomally normal boys. However, during puberty both left and right hemisphere functions along with testosterone levels and quality of parenting jointly predict the degree to which cases are free from a variety of psychopathologically defined characteristics.

We have attempted to explain these observations in terms of concepts offered by Tucker and Williamson (1984). They proposed

TABLE 6.1

Results of Stepwise Multiple Regression Analysis for 23 Extra X
Prepubertal Males

| Dependent Variable | Independent Variable | R2 Change | Partial r | F | p |
|---|---|---|---|---|---|
| Achievement | VIQ | .460 | -.678 | 17.92 | <.01 |
| Development | VIQ | .371 | -.609 | 12.41 | <.01 |
| | FSH | .107 | -.412 | 9.17 | <.01 |
| Family Relations | Parenting Quality | .707 | -.841 | 50.75 | <.01 |
| Delinquency | VIQ | .271 | -.520 | 7.80 | <.05 |
| Withdrawal | LH | .183 | -.427 | 4.69 | <.05 |
| Psychosis | VIQ Hand | .213 | -.462 | 5.70 | <.05 |
| Hyperactivity | LH | .351 | .561 | 9.67 | <.01 |
| | PIQ | .300 | -.662 | 16.00 | <.01 |
| | VIQ | .092 | .488 | 15.31 | <.01 |
| Social Skills | PIQ | .491 | -.491 | 6.66 | <.05 |
| Intellectual Screening | VIQ | .629 | -.792 | 35.57 | <.01 |
| | Hand | .069 | -.432 | 23.13 | <.01 |

Note. From "Relationships between hemispheric lateralization, sex hormones,
quality of parenting and adjustment in 47,XXY males prior to puberty" by C.
Netley, 1988, Journal of Child Psychology and Psychiatry, 29, 281-287.

**TABLE 6.2**

**Results of Stepwise Multiple Regression Analysis for 23 Extra X Pubertal Males**

| CBCL Dependent Variable Behaviour | Independent Variable | R2 Change | Partial r | F | p |
|---|---|---|---|---|---|
| Externalizing | Quality of parenting | .428 | -.653 | 15.69 | <.01 |
| | RH | .126 | -.469 | 12.41 | <.01 |
| | Testosterone | .137 | -.554 | 14.16 | <.01 |
| Schizoid | RH | .239 | -.498 | 6.59 | <.05 |
| Uncommunica- | | | | | |
| tiveness | Quality of parenting | .233 | -.482 | 6.38 | <.05 |
| | LH | .151 | -.443 | 6.23 | <.01 |
| | LuH | .109 | -.419 | 6.14 | <.01 |
| | RH | .130 | -.506 | 7.41 | <.01 |
| Immature | Testosterone | .305 | -.552 | 9.21 | <.01 |
| Obsess/Compul | Quality of parenting | .173 | -.416 | 4.40 | <.05 |
| | Testosterone | .191 | -.480 | 5.72 | <.05 |
| | PIQ | .123 | -.440 | 6.03 | <.01 |
| Hostile/ | Quality of parenting | .305 | -.552 | 9.20 | <.01 |
| Withdrawal | Testosterone | .253 | -.604 | 12.64 | <.01 |
| Delinquency | Quality of parenting | .530 | -.728 | 23.70 | <.01 |
| Aggressivity | Quality of parenting | .327 | -.572 | 10.22 | <.01 |
| | RH | .277 | -.514 | 10.21 | <.01 |
| | LuH | .136 | -.525 | 11.34 | <.01 |
| Hyperactivity | Quality of parenting | .309 | -.555 | 9.37 | <.01 |
| | Testosterone | .159 | -.480 | 8.80 | <.01 |
| | RH | .109 | -.453 | 8.65 | <.01 |
| | Estradiol | .093 | -.525 | 10.50 | <.01 |
| | LH | .081 | -.576 | 13.65 | <.01 |
| **Competence** | | | | | |
| Activity | VIQ | .389 | .623 | 13.35 | <.01 |
| Social | VIQ | .513 | .716 | 22.15 | <.01 |
| School | VIQ | .462 | .679 | 18.04 | <.01 |

Note: Only statistically significant results are presented.

Abbreviations: VIQ= Verbal I.Q., PIQ= Performance I.Q., FSH= follicle stimulating hormone, LH= Left hemispheric specialization, RH= Right hemispheric specialization, LuH= Luteinizing hormone.

that the left and right hemispheres have distinguishable roles in separate aspects of social and emotional behaviours. The left is primarily responsible for goal-related or motivational behaviour, while the right is concerned with responding to the environment physically and mediating affect. The findings just cited suggest that the inactive, unresponsive patterns of behaviour characterizing extra X boys before puberty are the result of impairments in left-hemisphere-based motivational systems. Data obtained somewhat later in development suggest that this system is modified after the onset of puberty by the introduction of right hemispheric processes which, when particularly strong, reduce the incidence and severity of maladjustment or psychopathology. Perhaps emotional flexibility or resiliency is enhanced in those 47,XXY males with particularly well-developed right hemispheric mechanisms, permitting them to better cope with the stresses associated with pubertal onset, testosterone deficits, and environmental adversity resulting from poor parenting.

## THE GENETIC-NEUROLOGICAL CONNECTION

If disorders of hemispheric specialization underlie various aspects of behavioural abnormality in 47,XXY boys, it is appropriate to ask what is responsible for the neurological anomalies themselves. Identifying the factors that give rise to hemispheric specialization is complex, as is determining whether hemispheric specialization is relatively fixed early in life or, alternatively, is emergent and changing from birth to maturity. Although these issues have not been resolved, various lines of evidence indicate that the two hemispheres have distinguishable properties early in life, suggesting that some elements of specialization may exist well before maturity.

This line of reasoning has been argued most cogently by Geschwind and Galaburda (1987) on the basis of hemispheric differences in neonatal anatomy (Wada, Clarke, & Hamm, 1975; Witelson & Pallie, 1973) that parallel those seen in mature individuals (Geschwind & Levitsky, 1968). Their suggestion that hemispheric specialization is largely fixed at birth is reinforced by other observations indicating that the two brain halves of young infants possess different information processing capabilities (Best, 1988; Molfese, Freeman, & Palermo, 1975), and that there are limits on the abilities of the hemispheres to assume "inappropriate" functions because of lateralized disease (Dennis & Kohn, 1974; Kohn & Dennis, 1974).

If hemispheric specialization is fixed at birth, what events operating during fetal life are responsible for disturbances in its development? Geshwind and Galaburda (1987) argued that elevations in levels of fetal testosterone (or particular sensitivities to its effects) result in anomalous patterns of hemispheric organization by retarding left hemispheric development, thereby leading to increased frequencies of such phenomena as left-handedness, language delay, and dyslexia. They also hypothesized that testosterone has similar effects on the development of the thymus, leading to the prediction that immunological disorders should be associated with left-handedness and language dysfunction. Since there is evidence that 47,XXY males show higher than normal rates of immunological disorders (Rhodes, Markham, Maxwell, & Monk-Jones, 1969), left-handedness (Netley & Rovet, 1982b; Theilgaard, 1984), and language disorder (Netley, 1986), it is appropriate to consider the Geschwind and Galaburda hypothesis as a possible explanation for this constellation of anomalies.

The simplest implication of this proposal is that testosterone levels or receptor sensitivities should be elevated in 47,XXYs during fetal life. Although the data necessary to examine this possibility are not available, there is no a priori reason to indicate that appropriate investigations would reveal such elevations. In fact, since testosterone levels have been reported to be lower than normal in extra X males in the perinatal state (Sorensen, Nielsen, Wohlert, Bennett, & Johnsen, 1981), it is more reasonable to suspect that they are lower than normal in the prenatal state as well.

If the testosterone hypothesis is an implausible one for dealing with the range of phenomena observed in the extra X state, it does possess some useful qualities as a prototypical explanation. For example, its emphasis on continuously distributed processes rather than discrete "on-off" ones is attractive since it is clear that individual differences in extra X males are no less than in chromosomally normal individuals (Netley, 1986). It is also appealing since it postulates a mechanism that translates one class of biological events (hormones) into another (neural maturation) that, in turn, could serve to mediate individual differences in behaviour. Finally, it directs attention to bio-behavioural systems that are operative very early in life, something that appears necessary when considering the consequences of a major genetic abnormality.

Our own theorizing on this issue was prompted initially by the reports of Waber (1976, 1977) that delays and accelerations in pubertal development were associated respectively with relatively high spatial and verbal ability. Since there is evidence that extra X males are delayed in bone age development (Stewart et al., 1979) and, more significantly, have slow mitotic cell division rates (Barlow, 1973; Polani, 1981), we speculated that individual differences in maturational variables may be associated with variations in intellectual functioning and hemispheric specialization. This indeed proved to be the case in relation to bone age (Netley & Rovet, 1982a). However, our most striking findings were that both verbal ability (Netley & Rovet, 1982a) and hemispheric specialization (Netley & Rovet, 1987) were related to variations in a prenatally determined dermatoglyphic measure, total finger ridge count (TFRC), frequently interpreted to be an index of prenatal rates of mitotic cell division (Mittwoch, 1973). The nature of these relationships suggests that a slow rate of prenatal growth leads to particularly strong right-hemispheric specialization that in turn inhibits the development of left-hemisphere-based verbal functions. This formulation shares a number of features with the Geschwind and Galaburda (1987) hypothesis in the sense that in both, prenatal events are postulated that induce variations in inter-hemispheric relationships and subsequent behavioural development.

Although our growth-related proposal is supported by the ancillary evidence that TFRCs are lower than normal in extra X states, several unanswered questions remain: (1) Why do extra X females, who also have low TFRCs, show a more general depression in intellectual function in contrast to the specific verbal deficit of extra X males? (2) Why do the verbal deficits of only some extra X males become less obvious as they mature and pass into adolescence? Before addressing these issues, it is necessary to describe some provisional observations we have collected on our 47,XXY boys when most were in puberty. The data were obtained by re-administering, at a mean age of 15.2 years, the sex laterality procedures originally carried out prior to puberty at a mean age of 11.1 years. As before, composite LH and RH scores were computed and compared to age-matched controls. Other data available to us were provided by radioimmunoassay determinations of levels of testosterone at age 15.

Analyses of the laterality data indicated that the mean hemispheric specialization variable of the extra X boys did not differ from that of age-matched XY boys. That they did differ prior to puberty implies, of course, that hemispheric functioning changed between these two points in time. We examined this issue in a series of analyses designed to explore the sources of individual differences in left and right hemispheric specialization during puberty and also the sources of changes in these parameters between the first and second assessments.

Initially, we examined whether subgroups of boys with both high and low TFRCs and testosterone levels at 15 (created by median splits) differed on the hemispheric specialization variables. An analysis of variance of these data produced no significant findings. We then examined, by a repeated measures of analysis of variance whether TFRCs and testosterone levels were related to changes in hemispheric functioning between 11.1 and 15.2 years. This yielded two significant results. The first was an interaction between TFRCs and the age at which right hemispheric specialization was assessed. These data are presented in Figure 6.1 and indicate that the low TFRC boys with initially high right hemispheric specialization scores showed a marked reduction in this parameter at 15.2 years. In contrast, the high TFRC cases with initially weak (in fact, weaker than expected) right hemispheric specialization increased to a point approximating that of the low TFRC group. This interaction, suggestive of two different kinds of normalization in the extra X boys, was extended by a second significant result: a three-way interaction between testosterone level, TFRC, and time of testing. These data are presented in Figure 6.2 and indicate that these normalizing changes in high and low TFRC groups occurred most clearly in cases where testosterone levels were relatively high.

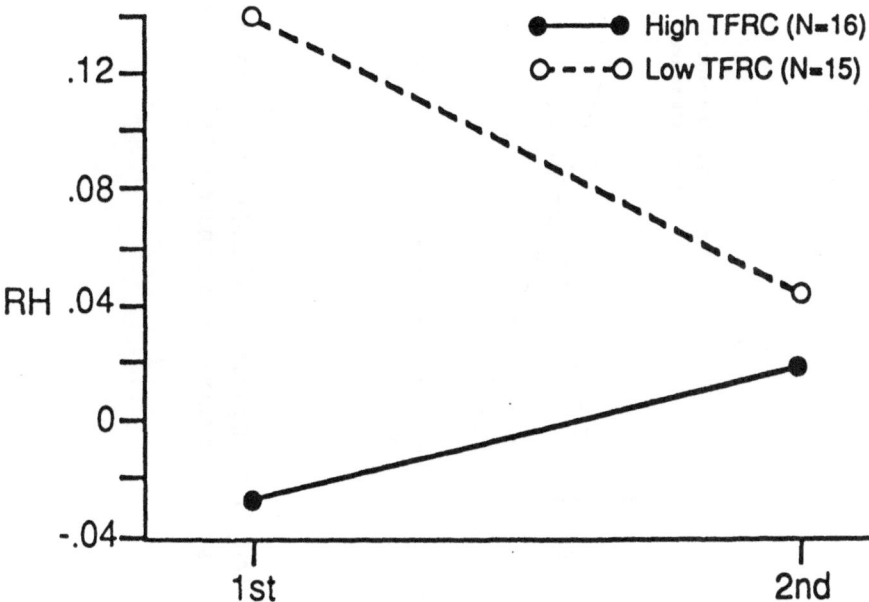

**Figure 6.1** Right hemispheric (RH) specialization during pre-puberty (1st) and puberty (2nd) in high and low TFRC (total finger ridge count) 47,XXY males.

## A MODEL OF NEURAL-BEHAVIOURAL DEVELOPMENT

If there is a changing system of neural development, how might it operate to account for the constant and variable features of behaviour shown by 47,XXY males from pre-puberty to adolescence? Furthermore, can such a model be extended to account for the behavioural anomalies shown by 47,XXX females?

In the case of 47,XXY males, most of whom have low TFRCs, it can be postulated that slow prenatal growth extends the period of right hemispheric maturational advantage described by Geschwind and Galaburda (1987), resulting in delayed development of the left hemisphere. A possible consequence of such a system would be that language acquisition in the left hemisphere would be delayed and may even shift to the right hemisphere. If inhibition of the less mature hemisphere by the more mature one is added to such a system, there would be a right-to-left inhibitory process that would further retard the development of the left hemisphere. This would be expected to result in a relative deficit in verbal abilities and, if the Tucker and Williamson (1984) proposal is valid, reductions in goal-directed motivational behaviours. In the case of the minority of extra X males

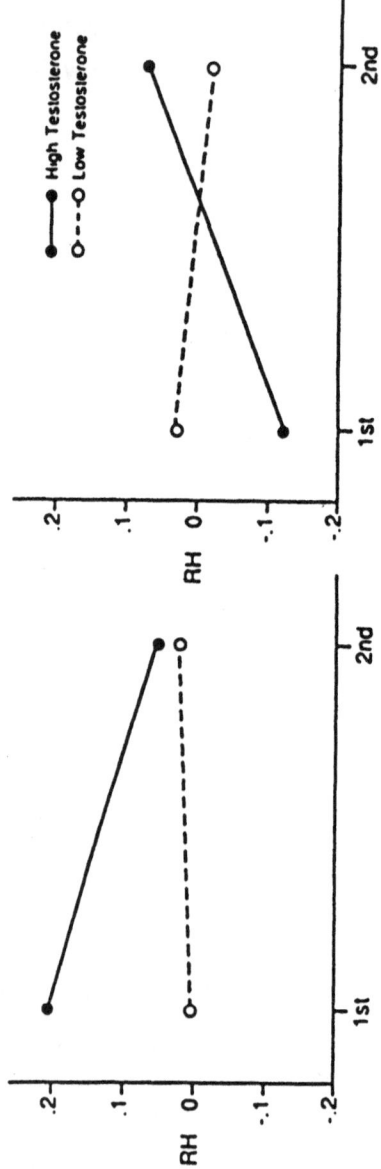

Figure 6.2 Right hemispheric (RH) specialization during pre-puberty (1st) and puberty (2nd) in 47,XXY males with high and low testosterone levels (taken at 15 years of age). (The low TFRC cases are presented in the left panel and the high TFRC cases in the right panel.)

with relatively high TFRCs, these events would be attenuated, permitting more normal development of both linguistic and temperamental functions.

What of puberty or, more particularly, of the effects of increasing testosterone levels associated with its emergence? There is evidence that low and high levels of testosterone during pubescence have different effects on the initial acceleration and the ultimate limitation of growth rate (Nielsen et al., 1986). This indicates that low levels of testosterone during early puberty increase growth, while higher levels later inhibit it. That is, the effects of testosterone may depend on the state of maturity or unrealized growth potential of biological systems subject to its influence. Therefore, it is possible that increased levels of testosterone during pubescence in 47,XXYs act differently on cases with greater or lesser degrees of right hemispheric maturation. Among those with well-developed right hemispheric functioning (the low TFRC subjects), this aspect of development would be inhibited by high levels of testosterone, whereas it would be stimulated among those with relatively immature right hemispheric maturation (high TFRC subjects). In the case of the former, right hemispheric inhibition would be attenuated providing the possibility of some reduction in the severity of the language deficit. The proposal implies that verbal ability would be relatively high in 47,XXY pubescent cases with high levels of testosterone. We have confirmed this prediction in an analysis of variance in which testosterone levels at age 15 were related to Verbal and Performance IQ in our extra X boys. It indicated that Verbal IQ was higher in high testosterone cases than in those with lower levels (means of 90.0 and 76.6, respectively, $p < .05$).

This formulation also implies that the social/emotional functioning of extra X males during adolescence might depend on levels of testosterone as well, since these would modify the hemispheric bases of emotional and motivational behaviour. The evidence presented earlier indicated that tendencies towards psychopathology were negatively related both to testosterone levels and right hemispheric functioning assessed prior to puberty. This represents something of a paradox since high testosterone (which reduces right hemispheric specialization) and strong right hemispheric specialization per se are both associated with relatively better adjustment. However, it is conceivable that strong right hemispheric specialization is useful for the promotion of good adjustment only when testosterone levels are low. In other words, it is possible that when testosterone approaches normal levels, 47,XXY boys with diminishing right hemispheric specialization are able to achieve relatively stable patterns of adjustment through adaptive systems that are like those of chromosomally normal males. More data are needed to evaluate this possibility.

The situation with 47,XXX females can be discussed only in terms of intellectual abilities since there are no clear data pointing to temperamental distinctiveness in these subjects. However, the available evidence on their intellectual functioning is reasonably consistent in indicating language deficits as well as a less pronounced depression of spatial ability (Netley, 1986).

It is known that as with extra X males, these phenotypically normal extra X females have a distinctive dermatoglyphic abnormality that results in low TFRCs. Simply extrapolating from the hypothesis presented earlier implies that they would have anomalies in hemispheric organization because of disturbances in prenatal growth rates. In particular, it would be expected that prior to puberty they would have relatively depressed left hemispheric functioning and relatively exaggerated right hemispheric functioning. Such a system could account for their relatively weak verbal ability but leaves unexplained their somewhat depressed spatial ability. To account for the latter, it appears necessary to postulate something in addition to a chromosomally-determined disorder in growth mediated by an anomaly in mitotic cell division. Testosterone could be such a factor, and borrowing from Geschwind and Galaburda (1987) it is possible that the absence of fetal androgens attenuates the delays in left hemisphere maturation, thereby providing less in the way of right hemispheric maturational advantage. Such a system would be more bilaterally equivalent, and any chromosomally determined slowing of growth would be expected to depress spatial functioning (although to a lesser extent than language abilities).

## CONCLUSIONS

Although the data presented in this chapter are clearly provisional in nature, they indicate that the origins of the behavioural distinctiveness of 47,XXY males are complex, and moreover, that they change during the course of development. Prior to puberty, hemispheric organization, which appears to be determined by prenatal growth and possibly hormonal events, is the best single predictor of the anomalies in language and temperamental functioning seen in these individuals. However, during puberty this relatively simple system is altered by the introduction of processes related to both hormonal and family functioning. Whether these factors continue to account jointly for individual differences in behaviour at maturity has yet to be determined. This essentially empirical question is only one of many suggested by the available data. Perhaps the most important of these is whether the findings and hypotheses presented in this chapter can be shown to be useful in understanding behavioural issues in other subject groups. These exist not only among X aneuploids such as 45,X and 47,XXX females (Netley, 1986), but also in chromosomally normal individuals with histories of prenatal or postnatal disturbances in growth or hormonal functioning.

## ACKNOWLEDGMENTS

The research described in this chapter has been supported by grants from the Ontario Mental Health Foundation and the Physicians' Services Incorporated Foundation.

## REFERENCES

Achenbach, T. M., & Edelbrock, S. S. (1981). Behavioral problems and competencies reported by parents of normal and disturbed children aged four through sixteen. Monographs of the Society for Research in Child Development, 46, 1-82.

Bancroft, J., Axworthy, D. I., & Ratcliffe, S. (1982). The personality and psychosocial development of boys with 47,XXY chromosome constitution. Journal of Child Psychology and Psychiatry, 23, 169-180.

Barlow, P. (1973). The influence of inactive chromosomes on human development. Humangenetik, 17, 105-136.

Bell, A. G., & Corey, P. N. (1974). A sex chromatin and Y-body survey of Toronto newborns. Canadian Journal of Genetics and Cytology, 16, 239-250.

Best, C. T. (1988). The emergence of cerebral asymmetries in early human development: A literature review and a neuroembryological model. In D. L. Molfese and S. J. Segalowitz (Eds.), Brain lateralization in children: Developmental implications, (pp. 5-34). New York: Guilford Press.

Dennis, M., & Kohn, B. (1974). Comprehension of syntax in infantile hemiplegics after cerebral hemidecortication: Left-hemisphere superiority. Brain and Language, 2, 472-482.

Geschwind, N., & Galaburda, A. M. (1987). Cerebral lateralization: Biological mechanisms, associations and pathology. Cambridge: The MIT Press.

Geschwind, N., & Levitsky, W. (1968). Human brain: Left-right asymmetries in temporal speech region. Science, 161, 186-187.

Kohn, B., & Dennis, M. (1974). Selective impairment in visuo-spatial abilities in infantile hemiplegics after cortical hemidecortication. Neuropsychologia, 12, 505-512.

Mittwoch, U. (1973). Genetics of sex differentiation. New York: Academic Press.

Molfese, D. L., Freeman, R., & Palermo, D. (1975). The ontogeny of brain lateralization for speech and nonspeech stimuli. Brain and Language, 2, 356-368.

Netley, C. (1986). Summary overview of behavioural development in individuals with neonatally identified X and Y aneuploidy. Birth Defects: Original Article Series, 22, 293-306.

Netley, C., & Rovet, J. (1982a). Verbal deficits in children with 47,XXY and 47,XXX karyotypes: A descriptive and experimental study. Brain and Language, 17, 58-72.

Netley, C., & Rovet, J. (1982b). Handedness in 47,XXY males. Lancet, 267.

Netley, C., & Rovet, J. (1984). Hemispheric lateralization in 47,XXY Klinefelter's syndrome boys. Brain and Cognition, 3, 10-18.

Netley, C., & Rovet, J. (1987). Relations between a dermatoglyphic measure, hemispheric specialization and intellectual abilities in 47,XXY males. Brain and Cognition, 6, 153-160.

Nielsen, C. T., Skakkeback, N. E., Darling, J. A. B., Hunter, W. M., Richardson, D. W., Jorgensen, M., & Keiding, N. (1986). Longitudinal study of testosterone and luteinizing hormone (LH) in relation to spermarche, pubic hair, height and sitting height in normal boys. Acta Endocrinologica, Suppl. 279, Vol. 113, 98-106.

Pennington, B., Bender, B., Puck, M., & Robinson, A., (1982). Learning disabilities in children with sex chromosome anomalies. Child Development, 53, 1182-1192.

Polani, P. E. (1981). Chromosomes and chromosomal mechanisms in the genesis of maldevelopment. In K. Connolly and H. Prechtl (Eds.), Maturation and development. Philadelphia: Lippincott.

Ratcliffe, S. G., Tierney, I., Nsahaho, J., Smith, L., Springbett, A., & Callan, S. (1982). The Edinburgh study of growth and development of children with sex chromosome abnormalities. Birth Defects: Original Articles Series, 18, 41-60.

Rhodes, K., Markham, R. L., Maxwell, P. M., & Monk-Jones, M. E. (1969). Immunoglobins and the X chromosome. British Medical Journal, 3, 439-441.

Robinson, A., Bender, B., Borelli, J., Puck, M., Salbenblatt, J., & Winter, J. (1986). Sex chromosomal aneuploidy: Prospective and longitudinal studies. Birth Defects: Original Article Series, 22, 23-71.

Sorensen, K., Nielsen, J., Wohlert, M., Bennett, P., & Johnsen, S. G. (1981). Serum testosterone with Karyotype 47,XXY (Klinefelter's syndrome) at birth. Lancet, 2, 1112-1113.

Stewart, D., Bailey, J. D., Netley, C., Rovet, J., & Park, E. (1986). Growth and development from early to mid adolescence of children with X and Y chromosome aneuploidy: The Toronto study. Birth Defects: Original Article Series, 22, 119-182.

Stewart, D., Netley, C., Bailey, J. D., Haka-Ikse, K., Platt, J., Holland, W., & Cripps, M. (1979). Growth and development of children with X and Y chromosome aneuploidy: A prospective study. Birth Defects: Original Article Series, 15, 75-114.

Theilgaard, A. (1984). A psychological study of the personalities of XYY and XXY men. Acta Psychiatrica Scandinavica Supplementum, 315, 1-131.

Tucker, D. M., & Williamson, P. A. (1984). Asymmetric neural control systems in human self regulation. Psychological Review, 91, 185-215.

Waber, D. P. (1976). Sex differences in cognition: A function of maturation rate? Science, 192a, 572-574.

Waber, D. P. (1977). Sex differences and the rate of physical growth. Developmental Psychology, 13, 29-38.

Wada, J. A., Clarke, R., & Hamm, A. (1975). Cerebral hemispheric asymmetry in humans. Archives of Neurology, 32, 239-246.

Walzer, S., Bashir, A., Graham, J. M., Silbert, A. R., Lange, N. T., DeHapoli, M. F., & Richmond, J. (1986). Behavioral development of boys with X chromosome aneuploidy: Impact of reactive style on the educational intervention for learning deficits. Birth Defects: Original Article Series, 22, 1-21.

Walzer, S., Graham, J. M., Bashir, A., & Silbert, A. R. (1982). Preliminary observations on language and learning in XXY boys. Birth Defects: Original Article Series, 18, 185-192.

144

Walzer, S., Wolff, P. H., Bowman, D., Silbert, A. R., Bashir, A., Gerald, P. S., & Richmond, J. (1978). A method for the longitudinal study of behavioral development in infants and children: The early development of XXY children. Journal of Child Psychology and Psychiatry, 19, 213-230.

Wirt, R. D., Seat, P. D., & Broen, W. E. (1977). Personality inventory for children. Los Angeles: Western Psychological Services.

Witelson, S. F., & Pallie, W. (1973). Left hemisphere specialization for language in the newborn. Brain, 96, 641-646.

# 7 Men with Sex Chromosome Aberrations— As Subjects and Human Beings[1]

## XYY AND XXY MEN AS SUBJECTS[2]

In the 1970s Witkin and collaborators initiated a case-finding study in Denmark on men with sex chromosome anomalies in the general population unselected with regard to institutionalization.

### Sampling Methods

The birth cohort consisted of all males born in the municipality of Copenhagen from January 1, 1944 to December 31, 1947, amounting to 31,438 men. The tallest 15% were selected for examination, since XYYs and XXYs tend to be more frequent among tall men (Noel & Revil, 1974; Witkin et al., 1976).

Information about height could be obtained in 28,884 cases. The losses were mainly due to death or emigration. A total of 4,591 men over 184 cm in height constituted 15.9% of the height distribution in the population of 28,884. Buccal smears and blood samples were collected from 4,139 (90.2%). For a detailed description of the case-finding procedure, see Witkin et al. (1976). The karyotyping procedures resulted in the identification of 12 XYY and 16 XXY men (Philip, Lundstrom, Owen, & Hirschhorn, 1976). All XYYs, 14 of the XXYs, and 52 matched XY controls agreed to participate in the individual case study.

### Selection of Controls

Two types of controls were assigned, designated "No BPP" and "BPP," respectively. The "No BPP" control was matched individually to the sex anomaly case for age, height, and social class. In addition to these three variables, the "BPP" control was also matched according to educational level and to scores on the army selection tests (The Borge Prien Prove, a Danish group intelligence test described in Rasch, 1960). Two groups of controls were formed in order to assess the effect of intellectual level (the "BPP" group), and to keep problems of overmatching under control (the "No BPP" group).

## Procedure

The data constitute the results of 16 hours of individual psychological and psychophysiological testing, the neurological examination, the two types of interviews conducted by the psychologist and the social worker, and information obtained from social, school, medical, and forensic records. This vast array of information was rated, scored, and coded by staff members, who had no knowledge of the genotypes of the subjects from whom the data were obtained. However, the exact number of XYY and XXY subjects in the sample was known to them. The scoring and evaluation of the case reports were made independently of the data from the cognitive tests and the interviews. During the two days of examination, all tests were presented in a standardized sequence and administered according to conventional procedures.

The selection of psychological tests. The psychological tests included cognitive measures of intelligence, memory, concentration, field dependence, and hemispheric lateralization, as well as projective measures of personality. Twenty-three tests in all were administered.

Psychological interview. Independent information on personality was obtained from a two-hour, semistructured interview. The essence of the answers could be rated quantitatively, and the questions were precoded. The ratings done independently by two psychologists were highly concordant. The interview also contained open-ended questions, and since the recording of the answers was verbatim, the outcome also lent itself to a qualitative processing of the material.

Social worker interview. The 59 precoded questions in this interview concerned educational and occupational development, social class, marital status, and parentage both of subject and biological (and possible substitute) parents. Details regarding number of residences, institutional placement, causes of job changes, the course of military service, and standard of present residence were also obtained for each subject. Finally, twenty-eight questions concerning present and previous health conditions were answered.

Hormonal assessment. Two blood samples were obtained on different mornings between 11 and 11:30 a.m. for determination of circulating testosterone, luteinizing hormone (LH), follicle-stimulating hormone (FSH), and prolactin levels. Significant high-reliability correlations were observed between hormonal values obtained from each subject. Analyses by thin-layer chromatography or urine samples were obtained, and drug influence thus checked. (For further details see Schiavi, Theilgaard, Owen, & White, 1984, 1988.)

Statistical procedures. The total statistical material consisting of approximately 61,000 data points was analyzed by D. R. Owen using the following tests: Cochran's Q-test, simple measure matched-groups ANOVA, and correlated measures matched-groups ANOVA.

## Results

Summarizing the quantitative findings comparing all the psychological data from the XYY and XXY men, it appears that the two groups are more alike than different. Thus the cognitive part of the study does not give evidence of different cognitive styles. Although both groups show a slight general deficit in global intelligence, a wide spectrum of IQ scores is possible within each. The IQ range was 77-124 in the XYY group, and 88-128 in the XXY group. However, it should be noted that a translation bias of a positive character (estimated to be around 10 IQ points) seems to influence the Danish version of the Weschler Adult Intelligence Scale (WAIS).

With regard to personality, the two groups of sex aneuploids also seem to have more rather than less characteristics in common. There is a slight difference concerning the defensive pattern, with the XYYs tending to be more evasive and the XXYs more inclined to use denial. The XYYs also seem more rigid in their thinking, while the XXYs are more indecisive. The latter group tends to be more submissive and dependent than the former, they are inclined to show less aggression against others, and criminal acts committed by XXYs tend to be less impulsive than those of XYYs.

The most distinct differences appearing between the two groups are related to the sexual domain. The XYYs are younger at first intercourse, have a higher frequency of masturbation both in childhood and adulthood (with less guilt feelings), report more libido and unconventional sex, and have more partners than the XXYs. They also show signs of a more differentiated sex role, are judged to be more masculine, and tend to have less problems with their masculine role than the XXYs. On the whole, however, the XYYs show a lower self-acceptance and their mood tends to be more pessimistic than that of the XXY men.

Finally, the ratings of the two sex aneuploid groups do not differ with regard to the major psychopathological diagnoses: psychotic, borderline psychotic, organicity (used here in the sense of signs of focal brain damage), somatization (tendency to react with psychosomatic symptoms), anxiety, and depression. None of these sex chromosome aberrations seems to present a specific or an exaggeratedly morbid picture.

Several questions are relevant. First, *Are there distinctive XYY and XXY syndromes of psychological features?* The picture emerging from this study suggests more similarities than differences between the groups with regard to cognitive and personality aspects. The diversities manifesting themselves belong mainly to the sexual domain: the gender role and the sexual behaviour. To a lesser extent, dissimilarity with regard to mediation of aggression is seen. If the term "syndrome" is applied in the strict sense, in signifying a characteristic dynamic combination or pattern of behaviour, then the degree and nature of the described disparities between the two karyotypic anomalies are not so essential that they warrant the description, "distinctive psychological syndromes."

Another hypothetical question is, *Do XXYs show more "feminine" and XYYs more "masculine" characteristics?* This naturally implies considerations concerning the psychological nature of men and women. Recent decades have witnessed a polemic debate characterized by polarization. An everyday experience, supported by numerous examples from experimental psychology, is that our conceptual framework influences our observations. This is especially valid when it concerns interpretations of complicated phenomena of which we have only an incomplete knowledge. Since the mid-1960s, investigations of biological as well as psychological sex differences have made rapid strides. With regard to the present study, the experimentally verified sex differences of a cognitive and emotional nature limit the relevant questions to the following:

1. Are XXYs superior to XYYs in verbal skills, and do these groups differ from their controls?
2. Are XYYs superior to XXYs on visuo-spatial tasks, and do these groups differ from their controls?
3. Are XYYs more field-independent compared to XXYs or to their controls? Are XXYs more field-dependent than their controls? Are there any differences with regard to lateralization?
4. Are XYYs more aggressive than XXYs? Do these groups show any differences compared to their controls?
5. Are there any differences in the acceptance and quality of sex role, when sex aneuploid males are compared among themselves and to their controls?

The factor analysis of the subtests in the WAIS, particularly the Verbal Comprehension vs. the Perceptual Organization factors, revealed no differences among XYYs, XXYs, and controls. Nor did an analysis of variance show any significant differences among the groups on the Masculinity-Femininity Scale.

A superiority in verbal vs. spatial tasks should manifest itself in a battery of memory tests differentiated with regard to material- and modality-specific factors. The results from these tests did not provide support for any such difference between XXYs, XYYs, and controls.

The segregation of functions as a result of psychological differentiation is likely to be reflected in the neuropsychological domain, and vice versa. Furthermore, to the extent that psychological functions are rooted in a neuropsychological substratum, specialization of functions at a neurophysiological level may actually be an important determinant of specialization of psychological function. From this it follows that field-independent persons, compared to field-dependent, should show greater lateralization of verbal processing in the left hemisphere and visuo-spatial processing in the right hemisphere. The XXYs show a very slight tendency to be more field dependent and less lateralized than the XYYs, but most striking are the similarities between the sex aneuploid groups on these two dimensions compared to the controls. Certainly, whether they are compared to XXYs or to their controls, the XYYs do not tend to follow a more definite masculine pattern with regard to either field dependence or lateralization.

When considering the various aspects of aggression as reflected in the projective battery and in the psychological interview, the differences between the groups are not of an impressive magnitude. Most noteworthy are variations in ways of manifesting aggression, rather than differences with regard to strength of aggression.

The fantasy productions of the entire projective battery yielded only two significant differences, both pertaining to the TAT: the XYYs have a higher aggressive-weighted sum than their controls ($p < .05$), but this is only valid when the results are unadjusted for vocabulary score; the XYYs have fewer anti-aggressive themes than the controls ($p < .01$). There are no differences between the sex aneuploid groups.

In the interview, several questions address various aspects of aggression. The self-reporting of the subjects discloses that, in the instances where manifest outer-directed aggression is concerned (e.g., having been a bully in school or showing violence against others at the present time), the XYYs do not differ significantly from either the XXYs or the controls. An exception is "aggression toward wife," where the XYYs differ markedly from their controls ($p < .001$), but marginally from the XXYs ($p < .07$). The information gained from questions pertaining to "relationship to partner" indicates that the XYYs generally have a less satisfactory alliance (including sexual relations) with their partner ($p < .05$). It is obviously difficult to tell cause from effect, but it does not seem farfetched to relate the aggressive attitude toward partners to sexual frustration and lack of self-acceptance, where XYYs also clearly differ from controls ($p < .05$ and $p < .01$, respectively). Although the XYYs do not differ from their controls in the case of verbal aggression, they do differ from the XXYs, who are significantly less sarcastic and teasing ($p < .05$). The latter group seems markedly more dependent and submissive ($p < .001$) both in comparison to their controls and the XYY group, which does not differ from its control group.

A pronounced characteristic of the XXYs is their anti-aggressiveness, which is reflected in the outcome of the TAT as well as in the answers to the interview. Furthermore, a correlational analysis of the various variables pertaining to aggression shows that while the XYYs follow the general trends indicated by the "No BPPs" with regard to covariation of parameters, many of the XXYs do not. This supports the impression gained from the individual case stories that aggressiveness in the XXYs takes on more disguised and defended forms, while in the XYYs it has a more direct, unpremeditated outlet. The XYYs consistently report more impulsivity concerning the circumstances of the offensive acts than either their controls ($p < .05$) or the XXYs ($p < .01$). The XYYs admit having been more frequently arrested (but not convicted) than their controls ($p < .05$). The comparison with the XXYs yielded nonsignificant differences.

Analysis of the relation between hormonal levels and records of criminal conviction revealed that among XYY men, testosterone levels were almost identical for subjects with and without convictions. However, the controls with convictions had significantly higher testosterone levels than the controls with no convictions. This finding suggests that aggressiveness might have a biological component associated with male sex hormones. Again it should be stressed that a correlational relationship does not necessarily reflect a causal one. More direct experimental proof is required before the significance of the hormonal factor in the interplay with behaviour and environmental factors can be delineated and more definite conclusions drawn. The complexity of the matter also requires that the question of aggression not be viewed in isolation. For example, that the XYYs report the quality of their childhood as poorer, in comparison to the controls and the XXYs ($p < .05$), underlines the intricate interactions of experiential, environmental, and genetic factors.

However, the activating and organizing role of the sex hormones should not be overlooked. The difference in the plasma concentrations of testosterone between XYYs and XXYs (Schiavi et al., 1978) undoubtedly shares a responsibility for the demonstrated dissimilarities in the libidinal strength and manifestations of sexuality among the two sex aneuploid groups.

In the Draw-a-Person test, the XYYs make the most differentiated drawings and the XXYs the least ($p < .05$). Qualitatively, the drawings of XXYs are generally less detailed and have inadequate proportions and poor integration. Sexual characteristics are either minimal or absent. The drawings of the XYYs show the opposite tendency, with overdimensionalized sexual attributes. This seems to mirror the fact that XXYs report significantly less libido than the XYYs, who also show a precocious development and a heightened sexual interest and activity as compared to their controls (Schiavi et al., 1988).

In the TAT the XXYs indicate more problems with their masculine role than the XYYs ($p < .01$). The subjective aspects of sex roles as expressed in the interview are also in agreement with these findings. The XXYs rate themselves lower than their controls on items pertaining to the masculine gender in childhood as well as adulthood. This does not imply that the XXYs have a feminine attitude. In fact, none appear to be homosexually oriented. However, they do seem to be more undifferentiated sexually than either the controls or the XYYs. The lack of a crystallized sense of masculinity in these men naturally also influences the attitudes of parents, siblings and/or peers.

That the XXYs have difficulty living up to the masculine role might be expected; it seems more surprising perhaps, that the XYYs also show a weakness of the sense of masculinity, as compared to the controls. For these men it is not so much a case of lack of sexual polarization toward the other sex as lowered self-acceptance and assertiveness. Another difference between the sex aneuploid groups seems to be that the less developed sense of gender is already marked in the XXYs from childhood, while the information from the interview suggests that the lack of a strong sense of masculinity in the XYYs first appears at puberty.

These results indicate that the attitude of a masculine gender is a subtle and complicated phenomenon. Its development seems not as dependent on an extra Y chromosome or an elevated testosterone level as on personality and environmental factors. The role of gonadal hormones in the case of aggressiveness is much more obscure. As pointed out by Meyer-Bahlburg (1974), evidence of the part played by the postnatal androgens in human aggression is still meager.

Another relevant question is, *Does the extra chromosome have an adverse effect on development?* A developmental lag may account for eventual differences in cognitive style between the two sex aneuploid groups and the controls. Subsidiary issues such as intellectual dysfunction, of a general or specific nature, and delayed or arrested hemispheric specialization also have to be taken into account. The developmental lag hypothesis also implies emotional immaturity and subordinate questions connected with instability and impulsivity.

Of the hemispheric lateralization tests, dichotic listening shows a difference between the XYYs and the controls, indicating that the left hemisphere is less completely lateralized in the former group. For the XXYs this is valid to an even higher degree when it concerns syllables; the processing of syllables is also more faulty in the right hemisphere compared to their controls. The hand preference scale accordingly shows more ambidexterity in the XXY group than in any of the other groups. The hand-tapping tests show greater discrepancy in hand proficiency between the sex aneuploid groups and their respective controls than does the hand preference test. This can be explained by the fact that hand preference is more subject to environmental influence.

Individuals who are less than fully right-handed have been found to be relatively more field dependent than clear right-handers (Pizzamiglio, 1974; Silverman, Adevai, & McGough, 1966). This represents another link between the two theoretical frames of reference: field dependence-independence and hemispheric lateralization.

As pointed out by Witkin and Goodenough (1981), the characteristic sequence in individual development is from a field-dependent to a field-independent mode of functioning. A developmental lag would thus imply that the field-dependent aspects of the cognitive style were dominating. Although the results from the Embedded Figures Test are in the direction of greater field dependence in both sex aneuploid groups, the differences are only statistically significant in comparison to the nonintelligence-matched controls.

Emotional immaturity may manifest itself in a number of ways: more intense and undifferentiated affects; impulsivity and lability; low threshold of frustration; dependency and lack of responsibility and social skills; retarded psychosexual development; and insufficient or more "archaic" defense mechanisms reflecting an immature ego-organization. The results of the projective tests, as well as the subjects' self reports, bear witness to immaturity and less differentiated and more labile affects than their controls. They also show less situative feeling and poorer judgment, and their personality integration is assessed to be more disharmonious than that of their controls.

Concentration capacity is generally poorer in the sex aneuploid groups than in the controls as measured by the color word naming test, the dichotic listening test, and self reports. The WAIS factor analysis shows results in the same direction on the Attention/Concentration variable, but nonsignificantly.

As the reduced inhibitory control over sensory function could be expected to result in distractibility, insufficient inhibitory control over motor function may result in restlessness and/or unfit motor activity. The XYYs report in the interview that they have more difficulties in relaxing than do the controls ($p < .05$). They are also more excitable and take more sedative drugs. Both sex aneuploid groups show significantly poorer motor performance in the hand-tapping test than the controls.

The impression that inhibitory control is less developed in the aneuploid than in the XY males also gains support from other data derived from the same sample as that of the present investigation. The Boisen and Rasmussen (1978) tremor study shows that physiological and intentional tremor are seen more often in XYYs than in controls, with a significant difference in the mean tremor amplitude ($p < .01$). The same tendency is reported in the XXYs, although not to the same extent. The results of the finger- and hand-tapping tests show significantly slower tapping by both aneuploid groups as compared to the controls.

Working with the present sample, Volavka, Mednick, Sergent, and Rasmussen (1977) focused on the electrocortical parameter and found that the XYYs have a significantly lower average frequency of the occipital alpha-activity than their controls. Both aneuploid groups showed significantly more theta-activity than their controls but it was largely within normal limits. It should be noted that the differences found were of a nonspecific nature. Volavka and co-workers see the differences as conceivably indicating retardation in the development of brain function. The fact that both of our aneuploid groups show significantly more theta activity than controls may be taken as an indication of a comparatively lower arousal. A preliminary analysis of the autonomic nervous system reactions of the XYYs in our study showed a low responsiveness and a slow rate of recovery of the skin conductance response.

These findings support the hypothesis that the extra chromosome has an adverse effect on development. Thus developmental lag is manifested in a more incomplete lateralization of the hemispheres. That the discrepancy between aneuploid and control groups generally is greatest with regard to left hemisphere processing may be interpreted as an indication of maturational arrest. The suggestion by Semmes (1968) that the left hemisphere is more specialized during development supports this view.

The neuropsychological findings in this study do not support the assumption that the XYY and XXY conditions cause focal brain damage. Rather, the results are consistent with the theoretical position that early alterations in brain development have generalized effects on brain function in later life (Luria, 1973). It should be stressed that it is not a question of grave brain damage, but rather of subtle disturbances that may lead to reduced tolerance of psychological and environmental strain.

The fact that the hormonal patterns of XYYs and XXYs differ significantly from those of the controls suggests that this is due to a combined effect of both neurological and neuroendocrine factors. However, our understanding of the fine mechanisms by which altered endocrine conditions affect the metabolism of the brain remains to be elucidated along with how these chemical changes influence the psychological pattern of such functions as arousal, activity, excitability, impulsivity, etc. Consequently, in our present state of knowledge, the interpretation of the role of the neuroendocrinological findings in the causal chain are highly speculative. The endocrine factors constitute only one of the potentials for influencing behaviour.

Psychosocial experiences may affect both the neural substrates and the consequent behaviour which these neural structures subserve, given the particular state of development. For example, that XYYs have spent more time in institutions and possibly have been exposed to more emotional deprivation should also be taken into account in the analysis of causal factors. The XYY and the XXY men are the result of an interplay of genetic, environmental, and psychological processes, as are all human beings. As Kessler and Moss (1970) point out, ascribing specific behavioural effects to a single chromosome fails to take into account a broad range of relevant environmental factors, and also ignores the fact that phenotypes result from the total integrative properties of genome-environment interactions.

## THE XYY AND THE XXY MEN AS HUMAN BEINGS

The statistically "average man" is an artifact. Even the following descriptions, limited to the conclusions of the projective test results, run the risk of fragmenting the reality by simplification. For obvious reasons of discretion, the information gained in the psychological and social interviews is not discussed. In spite of these limitations, I hope the descriptions will reflect the variations in and idiosyncratic features of the men in this study. In other words, the presentation of the following six cases is not an attempt to derive a generalization from particular examples. On the contrary, I have chosen three XYY men and three XXY men from the lower, medium, and upper intelligence range, and as will be evident, their intellectual level is not the sole difference among them.

Interpretations are based on the premise that personality is not compartmental. It should be regarded within a holistic, dynamic frame of thinking; that is, all psychological processes are conceived as interrelated and in constant interplay. Thus greater significance is ascribed to the configurations of scores than to the separate ones. In this way an attempt has been made to balance relevant and valid personality descriptions and the presentation of data in a brief survey.

The universal use of the Rorschach test makes detailed description superfluous. It will suffice to say that the scoring procedure is based on the system given by Rapaport, Gill, and Schafer (1968). The Thematic Apperception Test (TAT) (Murray, 1943) is also well known. From the conventional tests the following cards were chosen: 1, 3BM, 6BM, 7BM, 10, 13, and 16. These eight pictures were supplemented by two cards being a counterpart to card 18GF in the original series. Both pictures invite aggressive themes. The Word Association Test is an extended version of that described by Rapaport et al. (1968). Inevitably the Danish version cannot convey every alternative connotation of each English stimulus word, but it taps the same variety of ideational areas and conflicts.

The outcome of these projective tests was processed and described without knowledge of either the results of the cognitive tests or the content of the interviews.

## XYY A: WAIS Full-Scale IQ=88, Verbal IQ=78, Performance IQ=103

This is not a case of uncomplicated feeblemindedness. His perceptual organization appears to be more developed than his conceptual ability. His verbalization is halting and his word usage often faulty. He is not unimaginative and appears to have less trouble in mobilizing ideas than in developing a theme. He is easily aroused by aggressive and sexual stimuli. He has a low frustration threshold and is impulsive with poor judgment. Feelings of loneliness, insufficiency, and bodily shortcomings are marked. His personality structure is weakly integrated, the dominant defense mechanisms being denial and evasiveness.

## XYY B:  FSIQ=108, VIQ=104, PIQ=112

This is a man of average intelligence with his force in the practical field. His thinking is earthbound and labored, and he is very circumstantial. He is emotionally immature, apparently taking great pains to control the outward manifestation of his childish affects. In relationships with others he is rather taciturn and inhibited. Inner tension is probable, and sporadic sullen or dysphoric outburst and a vague inner unrest cannot be precluded. His defensive system seems disharmonious. Isolation, reaction formation, a certain stamp of compulsivity and evasiveness are most conspicuous among his defensive maneuvers.

## XYY C:  FSIQ=124, VIQ=125, PIQ=119

His intellectual abilities are above normal, and he will probably work accurately and cautiously without much zest or imagination. Intellectually inhibited and rigid, perhaps even compulsive, he is also very limited in his emotional register. He seems able to tolerate affects in only their most unobtrusive and adapted form. He is emotionally susceptible, and lacking the capacity to express his affects more freely and the potential to be introspective, the resulting intrapsychic tension is liable to cause somatic reactions. Inhibition, reaction formation, and rumination are the most dominant defense mechanisms. He seems to be a cautious, rigid, rather inhibited, anti-aggressive young man.

## XXY A:  FSIQ=92, VIQ=93, PIQ=91

The test results give the impression of a dull-normal intelligence. He tries hard, but his capacity for surveying the situation is weak, and he is obviously at his best when dealing with neutral, familiar, and clearly-defined practical tasks. Emotionally he is insecure and dependent, very set on adapting to his surroundings. He is rather sensitive, but his ability to adequately channel his affects is very limited; he is shy, quiet, and afraid of showing his emotions. This is also valid for aggressive feelings. There is a definite anti-aggressive feature in his personality, which is characterized by guilt feelings and mood swings in a depressive direction. His defense mechanisms are repression, reaction formation, rumination, and undoing.

## XXY B:  FSIQ=113, VIQ=122, PIQ=101

The passive, unimaginative stamp to his thinking, together with the indications of faulty perceptual organization makes it difficult to give an estimation of his intellectual level. Presumably he is fully normal, but it seems obvious that the cognitive style is developed very unevenly. He probably works satisfactorily in set tasks that present no problems and require no emotional engagement; he does not display much initiative. Emotionally he is somewhat immature, not spontaneous or straightforward in his way of relating. There is a heavy, inert stamp to his efforts. Anxiety is manifest, and there may be a tendency to a lowered mood. Repression and evasiveness seem most conspicuous.

## XXY C: FSIQ=119, VIQ=119, PIQ=116

Intellectually he is undoubtedly above normal, and his assets are within the imaginative field. However, he seems rather reluctant to realize his creative potential. It is difficult to decide alone from the test protocol whether he is habitually restrained and gloomy. He has strong feelings, but he inhibits the outlet of his affections. He probably withdraws somewhat from contact, at least to the extent of feeling lonely, and seeks compensation in fantasies and daydreams. His difficulties seem to lie mainly within the sexual and identity areas. The defense mechanisms are intellectualization and projection. His capacity for repression is not very developed, and a certain vulnerability is suggested.

## DISCUSSION OF THEORETICAL ASPECTS

Descartes dichotomized man into mind and body. His dualistic viewpoint has had a profound effect on Western thought, and until recently it has governed scientific methodology. In many ways the Cartesian approach has been successful, especially in biology. However, it has not only caused a split between body and mind by limiting the directions of scientific research, it has also made a division between the natural and the humanistic sciences. As physicists have gone beyond the Newtonian model, so it is also time for those working within biology and psychology to expand their epistemological views. Although it has been good practice in certain scientific quarters to claim the rule of parsimony with reference to theories and methods, an awareness of restrictions imposed by the framework is growing, such as the tendency to see man from a reductionistic, mechanical viewpoint. Principles of categorization determine that which is perceived. We see what we look for. Answers arising within a restrictive perspective tend to be curiously tangential to the original questions. There is a lack of coherence and meaning. It is as though the problems never quite permit themselves to be resolved according to the terms implied by the suggested solutions. Vital factors may thus be neglected. My argument does not indicate that the absolute truth, "Der Ding an sich" in Kantian terms, is to be found. But in matters as complicated as the interrelations of biological, psychological, and social factors, it is important to have several searchlights surveying the scene.

It does not enhance scientific thinking to try to make simplistic, straight-line cause-and-effect connections. When the human organism is viewed as a whole, involving interdependent physical and psychological patterns, it is necessary to attack the questions from different perspectives. One theory does not necessarily make another one redundant. The procedure of using both quantitative and qualitative methods does not inevitably lead to a cacophony of ideas or conceptual confusion. The two methods are complementary, in Niels Bohr's sense of the word, as are the disciplines of biology and psychology. Each of them calls for different categories of applied systems and variables. Psychology, in contrast to biochemistry, works with macroscopic data and large units, and therefore deals with deterministic and causal relationships within a phenomenological frame. It can hardly be expected that a conceptual frame referring to global phenomena will be valid on a microscopic level involving receptor cells, synapses, and transmitter substances. Stochastic models are likely to be preferred. Both quantitative and qualitative methods must be used in order to understand the interaction of genetic disposition, environmental conditions, and different kinds of trauma which are reflected in varying phenotypes. This can only be accomplished through a broad cross-disciplinary cooperation; otherwise the risk is an accumulation of data from a highly specialized area. Such data considered in isolation will give a fragmented picture, and does not further our understanding of the complex totality which is the object of our research: studying human beings in both their normal and psychopathological ways of functioning.

## CONCLUSIONS

It has been found necessary to make allowance for both idiographic-clinical and nomothetic-actuarial methods. The former mediates the individualized, global pictures of the personality, and may also serve to create ideas and formulate hypotheses. The latter may give information as to whether or not there is substance in the hypotheses.

It is now a generally accepted fact that no scientific approach is value-free or objective. This applies to both quantitative and qualitative methods. Although interpretation plays a part in both methods, the qualitative is the more subjective. To counteract results based on unsubstantiated intuition, the interpretations have been made using a double blind procedure. Furthermore, the psychograms were written before the outcome of the cognitive tests and information from the interviews were disclosed.

The different methods and the various sources of information serve as mutually controlling factors with regard to both reliability and validity. The outcome of the intercorrelations among test results and interview data, purporting to illustrate the same psychological factors, gives an impression of inner consistency.

158

Although these findings support the hypothesis that the extra chromosome has an adverse effect on development, it is important not only to acknowledge the genetic potential, but also to understand the special circumstances under which the genotype is operating. The chromosomal aberrations represent biological variants that leave the individual susceptible to a variety of endogenous and exogenous pressures. However, the problems are generally not of such magnitude that society needs to take special precautions. The variability of the phenotypes associated with sex chromosome aberrations should be stressed in genetic counseling. No single aspect in the interplay of factors has exclusive primacy. One should not anticipate that a person with a certain cytogenetic status will inevitably demonstrate an inflexible and irremediable personality pathology. The concept of a population at risk should be viewed with due regard to probability and multideterminism. As Shakespeare proclaimed, in King Lear, "Allow not nature more than nature needs."

## NOTES

1. The title indicates the two methods, the nomothetic and idiographic, applied in this psychological study of the personalities of noninstitutionalized XYY and XXY men.

2. This part of the investigation has been reported in detail elsewhere (Theilgaard, 1984, 1986), and is presented here in summary.

## ACKNOWLEDGMENTS

This investigation was supported by research grant (MH 23975) from the U.S. Public Health Service. The late Herman A. Witkin initiated the project of which this study is a part. Fini Schulsinger, M.D. from the Psychological Institute, Municipal Hospital of Copenhagen, and Raul C. Schiavi, M.D., Department of Psychiatry, Mount Sinai School of Medicine, New York also participated in aspects of this work.

## REFERENCES

Boisen, E., & Rasmussen, L. (1978). Tremor in XYY and XXY men. Acta Neurologica Scandinavica, 58, 66-73.

Kessler, S., & Moss, R. H. (1970). The XYY Karoytype and criminality: A review. Journal of Psychiatric Research, 7, 153-170.

Luria, A. R. (1973). The working brain. London: Penguin Press.

Meyer-Bahlburg, H. F. (1974). Aggression, androgens and the XYY syndrome. In R. C. Friedman, R. M. Richart, & L. O. Vande Wiels (Eds.), Sex differences in behavior. New York: Wiley and Sons.

Murray, H. (1943). Manual for thematic apperception test. Cambridge: Harvard University Press.

Noel, B., & Revil, D. (1974). Some personality perspectives of XYY individuals taken from the general population. Journal of Sex Research, 10, 3, 219-225.

Philip, J., Lundstrom, C., Owen, D., & Hirschhorn, K. (1976). The frequency of chromosome aberrations in tall men with special reference to 47,XYY and 47,XXY. American Journal of Human Genetics, 28, 404-411.

Pizzamiglio, L. (1974). Handedness, ear preference, and field dependence. Perceptual and Motor Skills, 38, 700-702.

Rapaport, D., Gill, M. M., & Schafer, R. (1968). Diagnostic psychological testing. R. Holt (Ed.). New York: International University Press.

Rasch, G. (1960). Probabilistic models for some intelligence and attainment tests. Copenhagen: Danish Institute for Educational Research.

Schiavi, R. C., Owen, D., Fogel, M., White, D., & Szechter, R. (1978). Pituitary gonadal function in XYY and XXY men identified in a population study. Clinical Endocrinology, 9, 233-239.

Schiavi, R. C., Theilgaard, A., Owen, D., & White, D. (1984). Sex chromosome anomalies, hormones and aggressivity. Archives of General Psychiatry, 41, 93-99.

Schiavi, R. C., Theilgaard, A., Owen, D., & White, D. (1988). Sex chromosome anomalies, hormones and sexuality. Archives of General Psychiatry, 45, 19-24.

Semmes, J. (1968). Hemispheric specialization: A possible clue to mechanism. Neuropsychologia, 6, 11-26.

Silverman, A. J., Adevai, G., & McGough, W. E. (1966). Some relationships between handedness and perception. Journal of Psychosomatic Research, 10, 151-158.

Theilgaard, A. (1984). A psychological study of the personalities of XYY- and XXY-men. Acta Psychiatrica Scandinavica. Suppl. 315, 69, 1-133.

Theilgaard, A. (1986). Psychologic Study of XYY and XXY Men. In S. Ratcliffe, & N. Paul (Eds.), Prospective studies of children with sex chromosome aneuploidy. New York: Alan Liss.

Volavka, J., Mednick, S. A., Sergent, J., & Rasmussen, L. (1977). Electroencephalogram of XYY and XXY men. British Journal of Psychiatry, 130, 43-47.

Witkin, H. A., Sarnoff, A. M., Schulsinger, F., Bakkestrom, E., Christiansen, K. D., Goodenough, D. R., Hirschhorn, K., Lundsteen, C., Owen, D., Philip, J., Rubin, D. B., & Stocking, M. (1976). Criminality in XYY and XXY men. Science, 193, 547-555.

Witkin, H. A., & Goodenough, D. R. (1981). Cognitive styles: Essence and origins. New York: International University Press.

Shirley G. Ratcliffe
Jennifer Jenkins
Peter Teague

# 8 Cognitive and Behavioural Development of the 47,XYY Child

Since 1965, attention has been concentrated on the behavioural implications of the 47,XYY genotype following the report of Jacobs, Brunton, Melville, Brittain, and McClement (1965) that 3.5% of the males in the Maximum Security Hospital at Carstairs, Scotland had this chromosome constitution. The literature relating to 47,XYY children changed its focus in 1968 from congenital malformation, mainly of the genitalia, to the association of tall stature and behavioural deviations as the indication for chromosome analysis. The inevitable result was that the association of these two features with the 47,XYY chromosome constitution was perceived to be very strong, despite the existence in the literature of a few reports of 47,XYY adults without disturbed behaviour (Cowling, Rigo, & Martin, 1969; Sandberg, Koepf, Ishihara, & Hauschka, 1961; Wiener & Sutherland, 1968). In the late 1960s, the need for population studies to establish the incidence of this and other chromosome abnormalities led to the inception of the newborn cytogenetic surveys in both the United States and the United Kingdom. These surveys provided cohorts of children whose development could be studied prospectively, free from the previous ascertainment biases.

In this chapter we present the results of one such study that includes the largest number of 47,XYY children identified at birth.

## MATERIALS AND METHODS

A cytogenetic survey of consecutive live born infants was conducted in two maternity hospitals in the United Kingdom from 1967 until 1979 (Buckton, et al., 1980; Jacobs, Melville, Ratcliffe, Keay, & Syme, 1974; Syme, 1974). A total of 34,380 infants were screened for sex chromosome abnormalities, either by chromosome analysis or by sex chromatin screening for abnormalities of the number of X and Y chromosomes, followed by full chromosome analysis if an abnormality was found. Among 17,522 males screened, 18 were found to have an additional Y chromosome in all cells, yielding an incidence of 1.03 per 1,000 male livebirths for the 47,XYY sex chromosome constitution. One of these infants died of renal agenesis shortly after birth. An additional 47,XYY infant was identified during a survey of all twin births at another hospital during the years 1969 to 1972.

Details of the study protocol and methods have been published elsewhere (Ratcliffe, Murray, & Teague, 1986). In that the ages of the eighteen 47,XYY boys comprising the entire cohort vary from 10 to 20 years, this report is limited to the 12 boys who were at least 13 years old at the time of data analysis. Three boys were members of dizygotic twin pairs, the co-twin being chromosomally normal (one of whom was female), and one set being identified in the twin survey referred to above. Of 110 chromosomally normal boys randomly recruited from the same newborn population, the 22 oldest served as controls and underwent identical assessment procedures, as did the co-twins.

The psychologist administering the tests was not informed of the karyotype of the child. Three sets of parents had been informed of the presence of an additional Y chromosome, in one case when the child was two months old and in the other two cases when the children were between 7 and 8 years of age.

## ANALYSIS AND RESULTS

Multiple linear regression was used to fit a model to the various psychological measures on each control child, taking into account social class (father's occupation) and mother's educational level. However, there were insufficient degrees of freedom among the cases to fit a regression model to both case and control data, as normally would have been done. Since there were more controls than cases, the significance of differences could have been overestimated, especially where the cases were more variable than controls. Consequently, predicted scores where calculated for the cases on the basis of the model obtained from the controls, but using each case's social class and mother's education score. A related $t$-test was used to compare these predictions with the observed values for each case, testing the hypothesis of no difference between controls and cases.

As the control boys were all singletons, scores of the 47,XYY singletons were compared with those of the controls; scores of the 47,XYY twins were compared with those of their chromosomally normal co-twins and will be presented separately.

## SOCIAL CLASS

The social class distribution of the cases and controls is given in Table 8.1 based on the occupation of the father (Registrar General's classification, 1970, professional I to unskilled V). It can be seen from this table that none of the fathers of the twelve 47,XYY children discussed in this report were in social class IV or V. This social class bias is not present when the entire case population of eighteen 47,XYY boys is examined, but it does exist in the sample that includes only the 12 oldest children. Hence, social class and

TABLE 8.1
Social Class Distribution (based on father's occupation)

| Social Class | 46.XY* | 47.XYY | |
| --- | --- | --- | --- |
| | | Singletons | Twins |
| I & II | 6 | 4 | 3 |
| III | 13 | 5 | 0 |
| IV & V | 3 | 0 | 0 |
| TOTAL | 22 | 9 | 3 |

*Controls

mother's educational level (Table 8.2) are controlled for in comparisons between the case and control populations.

## CLINICAL FINDINGS AT BIRTH AND DURING CHILDHOOD

As can be observed in Table 8.3, the birth order distributions of the case and control groups do not differ. Similarly, there were no differences in birth weight, supine length, or head circumference. However, the 47,XYY boys grew faster and were significantly taller than the controls by age 13 years ($p < .01$), with a mean height of 161.1 cm for singletons and 160.9 cm when the twins were included; both these values are between the 90th and 97th percentiles when compared with age and sex norms for British children (Tanner, Whitehouse, & Takaishi, 1966). The mean height of the control boys at age 13 (152.6 cm) was at the 50th percentile, and as illustrated in Figure 8.1, their distribution showed greater variability than that of the 47,XYY group.

The method of delivery at birth (Table 8.4) did not differ between the 47,XYY singletons and controls; however, all three 47,XYY twins were breech deliveries and were second born. The need for intervention immediately after birth is also shown in this table, with two infants in each of the case and control groups requiring assisted respiration. All case and control infants made satisfactory progress in the first week of life with none showing signs of cerebral trauma.

During childhood, one 47,XYY boy sustained a fractured skull followed by meningitis and subsequently developed severe unilateral hearing loss. One of the 22 control boys had a head injury for which he was hospitalized for 48 hours, but no fracture was demonstrated on X-ray of the skull. No convulsions were reported for any of the twelve 47,XYY boys nor for any of the 22 control boys.

All twelve 47,XYY boys were right-handed as were 20 of the 22 controls.

## SPEECH

In early childhood, speech development was delayed more frequently for the cases than the controls. Speech therapy was required by 33% (3/9) of the 47,XYY singletons and 18% (4/22) of the controls. One of the three case twins also required speech therapy.

## INTELLIGENCE TESTS

The Wechsler Intelligence Scale for Children (WISC) was administered at age 7 years, except in two cases where it was done at 11 and 12 years. The mean observed Verbal, Performance, and Full-

**TABLE 8.2**
**Mother's Educational Level**

| | 46XY | 47XYY Singletons | 47XYY Twins |
|---|---|---|---|
| University graduate | 1 | 0 | 0 |
| Other tertiary qualification | 2 | 2 | 1 |
| Advanced school leaving qualification | 3 | 2 | 0 |
| Ordinary school leaving qualification | 3 | 2 | 0 |
| No school leaving qualification | 13 | 4 | 2 |
| TOTAL | 22 | 10 | 3 |

**TABLE 8.3**
**Birth Order Distribution**

| Birth Order | 46XY | 47XYY | |
| --- | --- | --- | --- |
| | | Singletons | Twins |
| First | 12 | 5 | 2 |
| Second | 7 | 2 | 1 |
| Third | 2 | 2 | 0 |
| Fourth | 0 | 0 | 0 |
| Fifth | 1 | 0 | 0 |
| TOTAL | 22 | 9 | 3 |

**TABLE 8.4**
**Birth Events**

| Method of Delivery | 46.XY | | Singletons | 47.XYY | Twins |
|---|---|---|---|---|---|
| Spontaneous vertex | 18 | (82%) | 7 | (78%) | 0 |
| Forceps | 4 | | 1 | | 0 |
| Breech | 0 | | 1 | | 3 |
| Intervention | | | | | |
| Incubation and assisted respiration | 2 | | 0 | | 2 |

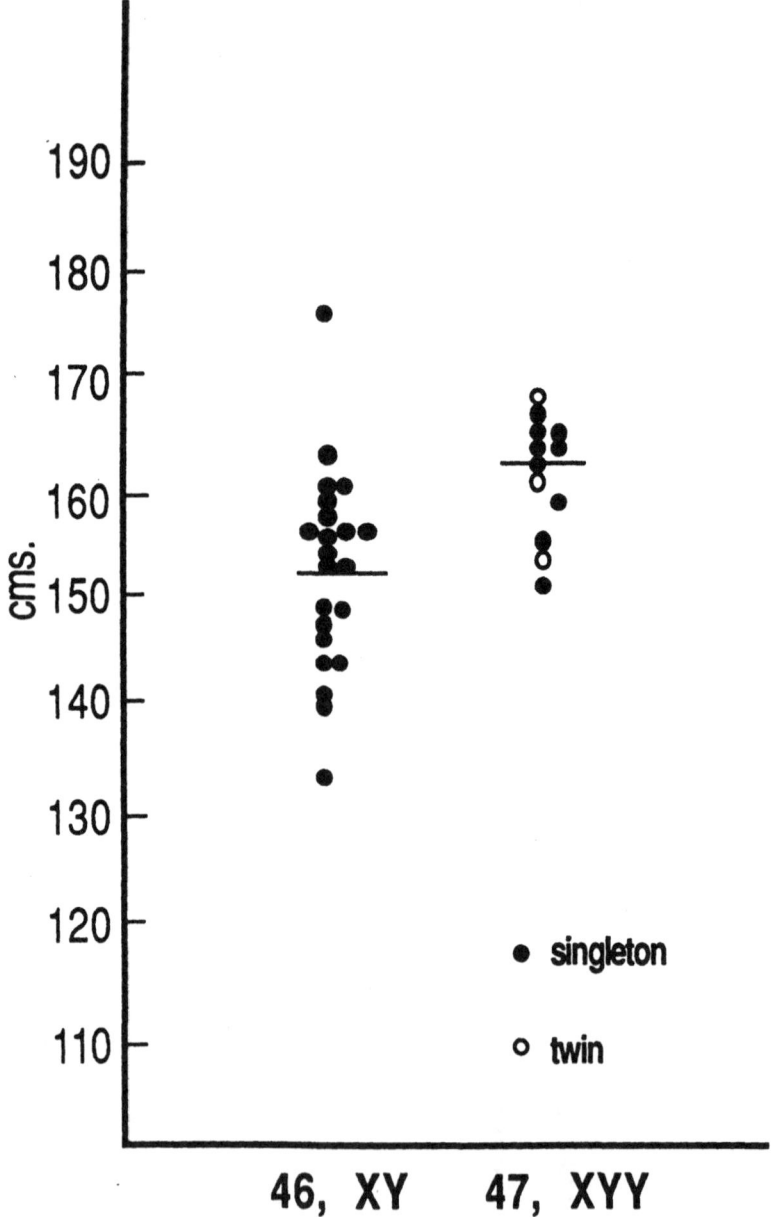

**Figure 8.1** Height of 47,XYY boys (both singletons and twins) and 46,XY control boys at age 13 years.

Scale IQ scores for the singleton 47,XYY boys were 100.2, 105.2, and 103.1, respectively; mean scores for the 47,XYY twins and co-twins (see Table 8.5) did not differ significantly from those of either the singleton 47,XYYs or controls.

Comparisons with predicted scores obtained from the multiple linear regression (allowing for social class and mother's educational level) revealed a small but statistically significant deficit for the 47,XYY boys in Verbal and Full-Scale IQ (both $p < .05$), but not Performance IQ. Two of the 47,XYY boys had Full-Scale IQ scores of at least 120, while the lowest score was 80.

## EDUCATIONAL ACHIEVEMENT AND BEHAVIOUR

### Numerical Ability

Numerical ability has been tested on eight case boys to date. When allowance was made for IQ scores, social class, and mother's educational level, there was no significant difference between the case and control boys.

### Reading Progress

Reading age was assessed initially at age eight years for the majority of children by means of the Burt Reading Test (Scottish Council for Research in Education [SCRE], 1976). When the child was tested at a later age an interpolated result was calculated. The 47,XYY boys exhibited reading scores ranging from 69 to 120 months. While two children were well above age expectation, six were below.

The mean observed reading score for the 47,XYY singletons was 93.4 months at a chronological age of 96 months. Controlling for social class and mother's educational level produced a predicted score of 100.7 months, from which the score obtained by the 47,XYY boys did not differ significantly. However, when the (lower) IQ scores of these boys were entered into the prediction equation, the score was reduced to 93.3 months, which was virtually identical to the obtained score. In other words, the 47,XYY boys as a group were reading in a manner expected for their level of cognitive ability and social class. Nevertheless, these results have to be viewed with caution. Remedial reading had been instituted by the teachers for four of the nine children in the case group (having no knowledge of their chromosomal abnormality), and four of the 22 controls, between 6 months and 2 years before the reading test was administered in the clinic. The boys receiving remedial help had lower reading scores than those who did not require this help.

Eight boys were tested again on word reading around 13 years of age using the British Ability Scales (Elliot, Murray, & Pearson, 1983). As can be observed in Table 8.6, there was no consistent change from their percentile positions at age eight. The most marked improvement

**TABLE 8.5**
WISC Scores for 47,XYY Boys (age 7 years)

### Singletons

|  | Verbal | Performance | Full-Scale |
|---|---|---|---|
| observed mean | 100.2 | 105.2 | 103.1 |
| predicted | 109.9 | 116.6 | 114.3 |
| t | 2.70 | 2.19 | 2.64 |
| p | <0.05 | <0.10 | <0.05 |

### Twins

#### 47.XYY

| Verbal | Performance | Full-Scale |
|---|---|---|
| 108 | 131 | 121 |
| 95 | 92 | 93 |
| 92 | 93 | 92 |
| Mean 98.3 | 105.3 | 102.0 |

#### Co-Twin

| Verbal | Performance | Full-Scale |
|---|---|---|
| 115 | 131 | 125 |
| 97 | 109 | 102 |
| 116 | 111 | 115 |
| Mean 109.3 | 117.0 | 114.0 |

**TABLE 8.6**
**Reading Progress in 47,XYY Boys**

| Age yrs-mos | Reading Age yrs-mos | Percentile (approximate) | Remedial Teaching | Age yrs-mos | Reading Age yrs-mos | Percentile |
|---|---|---|---|---|---|---|
| 9-0 | 9-6 | 40-50 | No | 15 | 12-6 | 32 |
| 10-0 | 6-5 | <10 | Yes | 14-10 | 8-9 | 6 |
| 8-0 | 8-3 | 40-50 | No | 14 | >14-5 | 70 |
| 8-7 | 7-1 | 10-20 | Yes | 13 | 10 | 22 |
| 8-0 | 7-6 | 20-30 | No | 13 | 11-7 | 28 |
| 8-0 | 7-7 | 30 | Yes | 13 | 13-1 | 50 |
| 8-0 | 7-8 | 30 | No | 13 | 10-5 | 20 |
| 9-0 | 7-9 | 24 | No | 13 | 12-9 | 42 |
| 9-5 | 6-7 | 1-2 | Yes | 13-3 | 13-8 | 53 |

was seen in the 47,XYY boy with the severest deficit, following his attendance over a three-year period at a special unit for dyslexic children.

## Behaviour in School

Behaviour in school was examined using the Bristol Social Adjustment Guide (Scott, 1976), completed by the teachers annually when the child was between the ages of 8 and 11 years. As this questionnaire is used mainly in the United Kingdom, a brief description is provided here. Under headings such as "interaction with teacher," "school work," "attitudes to other children," and "personal ways," the teacher has a choice of 5 to 6 behaviours from which to select what best describes the child (including "nothing noticed"). This results in 154 statements that are then coded in terms of core syndromes within the three main categories of under-reaction, over-reaction, and neurological. Under-reaction contains the four core syndromes, "unforthcoming, withdrawal, depression, and non-syndromic under-reaction," while over-reaction is made up of core syndromes labeled, "inconsequence, hostility (including aggressive behaviour), peer maladaptiveness, and non-syndromic over-reaction." The neurological score contains the following descriptions of behaviour: gets confused and tongue-tied; too restless and overactive to heed correction; very jumpy and easily scared; makes aimless movements with hands; has unwilled twitches and jerks. The neurological index is probably more accurately described as a score of anxiety.

The mean scores obtained in each category by the 47,XYY boys and controls are illustrated in Figure 8.2. As these scores were not normally distributed, a logarithmic transformation was used in the analysis. The mean observed score for under-reaction for the 47,XYY boys was higher than that of the controls ($p < .01$) both before and after controlling for social class, mother's educational level, and Full-Scale IQ. The case group also had significantly higher neurological scores than the controls ($p < .02$), but the level of over-reaction did not differ between the groups. In particular, there was no evidence of increased aggressive behaviour in the 47,XYY boys. The twin data, also shown in Figure 8.2 (but not included in the analysis), differ from the singletons' results in the category of over-reaction, with high scores in all three cases; however, these high scores were not reflective of higher levels of aggression.

## Behaviour at Home

Preschool behaviour was assessed at 3 to 3.5 years of age using the Behaviour Screening Questionnaire (Richman & Graham, 1971). This procedure consisted of mothers' reports of their child's behaviour in the areas of eating, soiling, sleeping, activity level, concentration, relationships, dependency, difficult behaviours, temper tantrums,

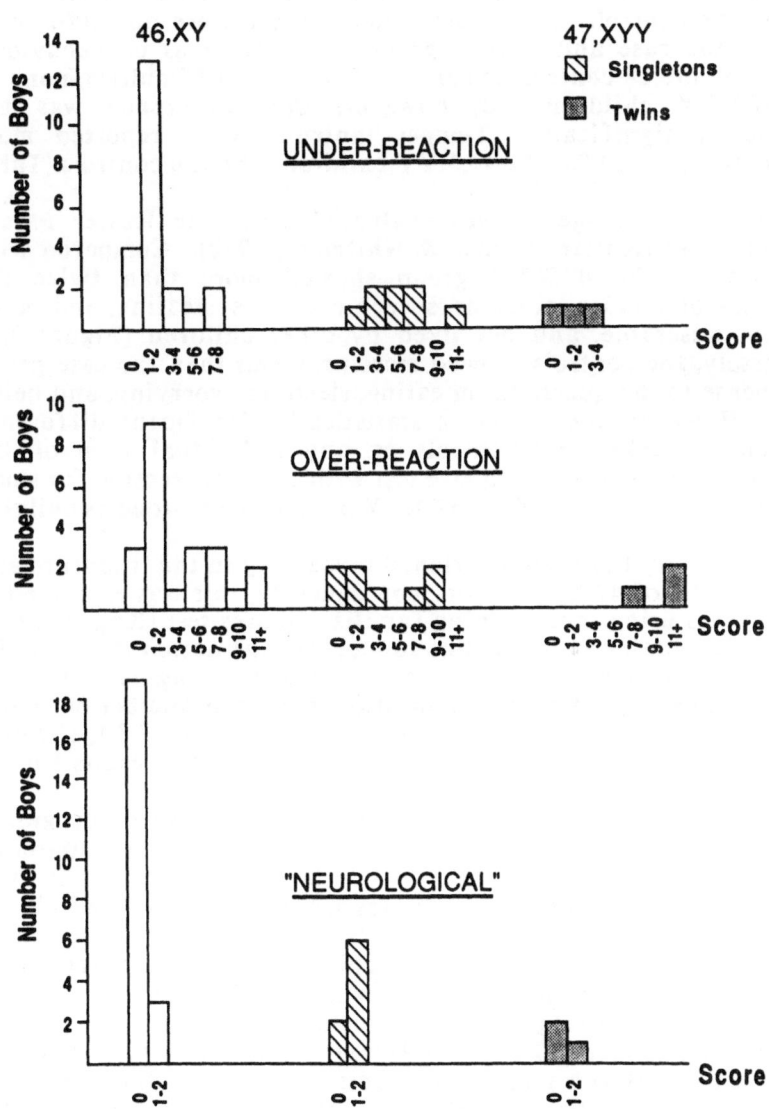

**Figure 8.2** Frequency distributions of mean scores in the three main categories of the Bristol Social Adjustment Guide obtained for 47,XYY boys (both singletons and twins) and 46,XY control boys between 8 and 11 years of age.

negative mood, worries, and fears. Results were available for 18 controls and 8 of the 9 case singletons. The overall mean score obtained by the 47,XYY boys (3.5) did not differ significantly from that of the controls (3.1). Furthermore, there were no differences between the case and control groups in most areas of behaviour. While 6 of the 18 controls experienced sleeping difficulties, none of the 47,XYY children did; however, this difference was not statistically significant. Temper tantrums were reported more frequently ($p < .02$) for the 47,XYY children than the controls (Table 8.7).

Behaviour at age 12 was evaluated using the Rutter Parent Questionnaire (Rutter, Tizard, & Whitmore, 1970). Compared with the controls, the 47,XYY group showed more than twice the frequency of temper tantrums, speech problems, stealing, and being solitary, miserable, and not liked by other children (Figure 8.3). Conversely, the controls showed more problems than the case group in response to the questions on eating, sleeping, worrying, and being fussy. However, there was no statistically significant difference between the cases and controls on any individual item of the questionnaire, using a chi-square test with Yates correction for small numbers. Finally, none of the 47,XYY boys were recorded as bullying other children.

Impulsivity has been mentioned repeatedly in the literature as a characteristic of 47,XYY adults; consequently, this was examined at age 11 years using the matching familiar figures test (Kagan, 1965). The results obtained provided no support for the notion that 47,XYY boys are impulsive (the time to correct choice being 11.16 seconds where 10.25 was predicted). In addition, the mean number of errors, 13.9, did not differ significantly from the prediction of 8.4, allowing for IQ, social class, and mother's educational level. Results for the three twins were in line with those for singletons.

In the development of child psychiatric disturbance, recognized risk factors include marital disharmony/breakdown and parental psychiatric illness (Rutter et al., 1975). The occurrence rates of these factors for the case and control groups are shown in Table 8.8. It can be seen that children in the case group experienced separation/divorce and maternal psychiatric illness three times more frequently than the control children, whereas there was no difference in the rate of paternal psychiatric illness.

Psychiatric referral (Table 8.9) was five times more frequent in the 47,XYY children than in the controls. None of the parents of referred case boys were aware of the chromosome abnormality at the time of referral; however, since the pediatrician (Ratcliffe) was, it is possible that this knowledge may have lowered her threshold for referring to a psychiatrist when a behaviour problem was reported repeatedly by the parent. Presenting complaints included persistent temper tantrums and difficult, demanding behaviour in all cases, stealing in two cases (with police involvement), and one each with enuresis, early morning weeping, and uncontrollable facial and body

**TABLE 8.7**
**Behaviour Screening Questionnaire Results**

|  | 46,XY | 47,XYY Singletons |  |
|---|---|---|---|
| **Sleeping Difficulties** |  |  |  |
| Present | 6 | 0 | N.S. |
| Absent | 12 | 8 |  |
| **Temper Tantrums** |  |  |  |
| Present | 3 | 6 | p<0.02 |
| Absent | 15 | 2 |  |

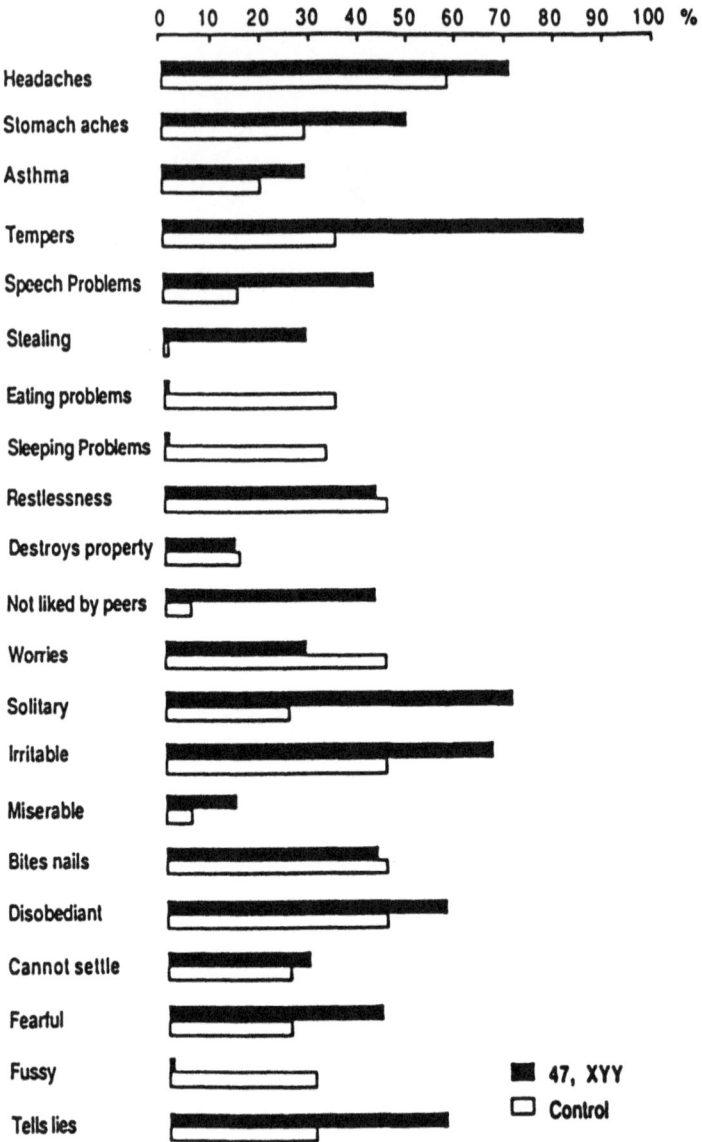

**Figure 8.3** Percentage of boys in the 47,XYY and control groups reported as exhibiting the behaviours (at age 12 years) described in the Rutter Parent Questionnaire.

**TABLE 8.8**
Incidence of Risk Factors

| | 46,XY | | 47,XXY | | |
| --- | --- | --- | --- | --- | --- |
| | | | Singletons | Twins | Combined |
| Parental separation/Divorce | 2 | ( 9.1%) | 1 | 1 | (33.3%) |
| Maternal psychiatric illness | 4 | (18.2%) | 4 | 3 | (58.3%) |
| Paternal psychiatric illness | 3 | (13.6%) | 1 | 1 | (16.7%) |

**TABLE 8.9**
Incidence of Family Pathology & Psychiatric Referral

| | 46,XY | | 47,XYY | | |
|---|---|---|---|---|---|
| | | | Singletons | Twins | Combined |
| Family pathology | 6 | (27.3%) | 5 | 3 | (66.7%) |
| Psychiatric referral | 2 | ( 9.1%) | 4 | 2 | (50.0%) |

movements when stressed. Two of the six referred children had learning difficulties, but none had experienced delayed speech development.

Therapeutic measures included counseling and pharmacological agents where appropriate. For example, one boy with severe temper tantrums and a posterior occipito-temporal focus on his electroencephalogram received carbamazepine; imipramine and desipramine were prescribed for two boys judged to have depressive reactions to environmental stress; another boy who was hyperactive was given methylphenidate. The symptoms resolved gradually in all children, either in response to the psychiatric intervention or with the progress of individual maturation, and none are currently attending psychiatric departments. Four of these cases have been reported in detail (Ratcliffe & Field, 1982). Of the controls, two were referred to their local educational psychologist because of stealing, truancy, and difficult behaviour.

Rates of behaviour problems in the case and control children were examined as a function of family pathology, the latter being defined as marital breakdown and/or parental psychiatric illness. Behavioural problems were assessed using the Rutter Parent Questionnaire, which contains items relating to health problems and antisocial and neurotic behaviors. Children with a score of 13 or higher were designated as having a behaviour problem. In both the control and case groups, children who experienced family pathology were no more likely to show behavioural problems as defined by a raised score than those who had not had this experience. These apparently anomalous findings may result from small numbers and/or the intended use of the Rutter questionnaire as a large scale screening tool rather than as an individual diagnostic measure.

By contrast, when psychiatric referral was used as the outcome measure there was a significant association between the presence of family pathology and psychiatric referral in the 47,XYY boys (Fisher's exact test, $p < .05$ for singletons), but no such association for the control children. Whether the higher level of family pathology in the cases occurred independently or secondarily to the presence of a 47,XYY child in the family cannot be determined at present.


## DISCUSSION

At this point in the study, conclusions remain preliminary and limited by small numbers, although this latter reservation will be overcome to some extent by the time the whole cohort of 18 boys have completed their development. Our data on speech development confirm an increased prevalence of delay in the group, and the results of IQ tests show a small deficit when controlled for social class. The range of IQ scores obtained was as wide as in the normal population. Furthermore, the presence of an additional Y chromosome does not preclude the achievement of superior scores, as two boys had Full-

Scale IQ scores of 120 or more (although in one case this was heavily weighted by a Performance score 23 points above his Verbal score). While the mean Verbal IQ of the 47,XYY boys was lower than their mean Performance IQ, the difference was statistically significant in only three instances (Ratcliffe et al., 1986). The intelligence-test results of our study are shown together with those of the other newborn surveys in Table 8.10. The scores are generally comparable, except for the three boys from the New Haven survey who scored lower (Leonard & Sparrow, 1986). However, from the description of these boys, this may have been the consequence of cerebral trauma at birth in two of the three cases.

In the cases reported here there was no evidence of birth trauma in the singletons. In the three twin 47,XYY boys, delivery was breech, which is more likely to be associated with trauma to the brain due to the lost opportunity of gradual molding of the head at the pelvic outlet by contractions during labor. However, these particular boys showed no clinical evidence of cerebral trauma in their behaviour during the newborn period, and their IQ scores were very similar to those of the singleton 47,XYY boys. During childhood, one of the 47,XYY boys sustained a cerebral insult in the form of a skull fracture with subsequent meningitis and unilateral hearing loss; while this may have been a significant episode in his personal development and behaviour, the effect of his IQ scores on the mean IQ of the case group was minimal, as he achieved a Verbal score of 95 and a Performance score of 93. IQ comparisons with siblings (Ratcliffe et al., 1986) were disadvantageous to all but one of the 47,XYY boys whose sibling was mentally retarded.

In the absence of other known significant factors affecting cognitive development in these boys, we conclude that the presence of an additional Y chromosome had a small deleterious effect on cognitive ability, although the mechanism by which this occurred remains speculative. The difficulty that the 47,XYY boys had in learning to read was substantial when viewed from the teachers' assessment that remedial help was needed in over 40% of the cases. However, it should be noted that for some of the children the remedial teaching was initiated before the reading test was administered. At this point we cannot exclude the possibility that the case children have a specific learning difficulty. The 47,XYY boys' achievements also may have been affected by their behaviour, which the teachers recorded to be high in terms of under-reaction or withdrawal. The high neurological score also raises the question of increased anxiety, which may have inhibited achievement of their full potential. The 47,XYY boys with lower reading ability maintained their percentile positions between ages 8 and 13 years with the aid of remedial help, although the worst affected boy, who attended a dyslexic unit for three years, showed impressive improvement in his reading ability.

The increased height of the 47,XYY boys may have led to higher expectations of them relative to their peers, making their slightly

**TABLE 8.10**
**Mean IQ Scores For 47,XYY Boys From Newborn Surveys (ages 7-13 years)**

| Authors | Number | Test | Verbal | Performance | Full-Scale |
|---|---|---|---|---|---|
| Robinson (1986) | 4 | WISC-R | 94.0 | 111.5 | 101.8 |
| Stewart (1986) | 3 | WISC-R | 92.3 | 92.6 | 91.3 |
| Evans (1986) | 3 | WISC-R | 100.7 | 94.7 | 97.7 |
| Leonard (1986) | 3 | WISC-R | 81.0 | 72.3 | 75.7 |
| Nielsen (1982) | 5* | WISC | 99.6 | 108.4 | -- |
| Ratcliffe | 12 | WISC | 99.7 | 105.2 | 102.8 |

*includes one mosaic 46,XY/47,XYY.

lower level of cognitive functioning more salient. However, there appeared to be no close association between increased height and development of disturbed behaviour, as the boys with the most marked difficulties who were referred for psychiatric help were at the 25th, 50th, 75th, 90th (2), and 97th height percentiles.

With regard to the behaviour of the 47,XYY boys, the use of behavioural questionnaires in the preschool period and in late childhood did not reveal an increase in total scores. Only when individual items of behaviour were examined did differences appear, with an excess of temper tantrums in the preschool period as well as in late childhood. However, the incidence for the 47,XYY boys (86%) at the latter time did not differ significantly from that of the controls (35%). At this stage in the study small numbers constitute a major problem in data analysis, but the questionnaires are nevertheless useful in illustrating a number of areas in which the behaviour of 47,XYY boys did not differ from that of the controls. Reports of sleeping and eating indicated that the 47,XYY boys were easier to handle than controls in both early and late childhood.

The higher level of family pathology for the case group than for the controls (65% vs. 27%) has to be viewed bi-directionally. It may have contributed to behaviour disturbance in the 47,XYY boys, but conversely, the behaviour of these children may have exacerbated problems between the parents. When the divorce rate was examined in the study as a whole, it was found to be 19% among the parents of 36 children with an extra X chromosome compared to 50% among the families of boys with an extra Y chromosome, lending some support to the suggestion that the more difficult behaviour of the 47,XYY boys contributed to family breakdown.

It is noteworthy that the mothers of the 47,XYY boys experienced three times the level of psychiatric illness occurring in the control mothers, where no difference was found in the rates for fathers. Marked social deprivation was present in only one control family, and hence was not a significant factor in family breakdown or behaviour disturbance. In addition, available ancillary services such as speech therapy, remedial education, and/or psychiatric help were sought by 11 of the twelve 47,XYY boys, but only by 7 of 22 controls. Where such support is not readily available, the outcome may be less favorable.

Descriptions of the behaviour of the 47,XYY boys by parents and teachers indicated that aggression was not characteristic of these children, nor was impulsivity demonstrated in the test situation. However, temper tantrums were frequent at both 3 and 12 years of age, and were a prominent feature in those children who were referred for psychiatric help. Temper tantrums should not be viewed as an expression of aggressive behaviour, but rather as an uncontrolled response to frustration and/or the result of a lowered threshold to frustration. Family instability was clearly an important factor associated with behaviour disturbance, as in chromosomally normal boys. However, improvement occurred in response to counseling, environmental modifications, and in appropriate cases, to pharmacological measures, emphasizing the interactive nature of the genetic constitution with the environment. While the uptake of support in terms of speech therapy, remedial education, and psychiatric help was higher than in controls, most of the children led normal lives with their families and attended ordinary schools.

One of the major objectives of a study such as this is to provide accurate information to parents faced with the decision whether to continue a pregnancy with an 47,XYY fetus. At this stage in our longitudinal study, we feel that the nature and magnitude of the problems encountered do not seem to be of such severity as to warrant recommending termination.

## REFERENCES

Buckton, K. E., O'Riordan, M. L., Ratcliffe, S. G., Slight, J., Mitchell, M., McBeath, S., Keay, A. J., Barr, D., & Short, M. (1980). A G-band study of chromosome in liveborn infants. Annals of Human Genetics, 43, 227-239.

Cowling, D. C., Rigo, S., & Martin, F. I. R. (1969). The "YY Syndrome". The Medical Journal of Australia, 2, 443-445.

Elliot, C. D., Murray, D. J., & Pearson, L. S. (1983). The British Ability Scales. Revised Edition. Windsor NFER.

Jacobs, P. A., Brunton, M., Melville, M. M., Brittain, R. P., & McClement, W. F. (1965). Aggressive behaviour, mental subnormality and the XYY male. Nature, 208, 1351-1354.

Jacobs, P. A., Melville, M., Ratcliffe, S. G., Keay, A. J., & Syme, J. (1974). A cytogenetic survey of 11,680 newborn infants. Annals of Human Genetics, 37, 359-376.

Kagan, J. (1965). Reflection-impulsivity and reading ability in primary grade children. Child Development, 36 609-628.

Leonard, M. F., & Sparrow, S. (1986). Prospective study of development of children with sex chromosome anomalies: New Haven study IV. Adolescence. In S. G. Ratcliffe & N. Paul (Eds.), Prospective studies on children with sex chromosome aneuploidy. Birth Defects: Original Article Series, 22, 221-249.

Ratcliffe, S. G., & Field, M. A. S. (1982). Emotional disorder in XYY children: Four case reports. Journal of Child Psychology and Psychiatry, 23, 401-406.

Ratcliffe, S. G., Murray, L., & Teague, P. (1986). Edinburgh study of growth and development of children with sex chromosome abnormalities III. Birth Defects: Original Article Series, 22, 73-118.

Richman, N., & Graham, P. J. (1971). A behavioural screening questionnaire for use with three year old children: Preliminary findings. Journal of Child Psychology and Psychiatry, 12, 5-33.

Rutter, M., Tizard, J., & Whitmore, K. (1970). Education, Health and Behaviour. London: Longmans Green.

Rutter, M., Yule, B., Quinton, D., Rowlands, O., Yule, W., & Berger, M. (1975). Attainment and adjustment in two geographical areas III. Some factors accounting for area differences. British Journal of Psychiatry, 126, 520-533.

Sandberg, A. A., Koepf, G. F., Ishihara, T., & Hauschka, T. S. (1961). An XYY Human Male. Lancet II, 488.

Scottish Council for Research in Education. (1976). The Burt Word Reading Test, 1974, Revision. London: Hoddler and Stughton.

Stott, D. H. (1976). Bristol Social Adjustment Guides, 5th Ed. London: Hoddler and Stughton.

Tanner, J. M., Whitehouse, R. H., & Takaishi, M. (1966). Standards from Birth to Maturity for Height, Weight, Height Velocity and Weight Velocity: British Children 1965. Archives of Disease in Childhood, 41, 454-471.

Wiener, S., & Sutherland, G. (1968). A normal XYY man. Lancet, 2, 1352.

*Daniel B. Berch*

# 9    Methodological Issues in SCA Research

The study of individuals with sex chromosome abnormalities (SCA) presents one of the most intriguing yet methodologically challenging research endeavors that a behavioral scientist can undertake. From initial selection of participants to the final interpretation of results, even the most experienced investigator is confronted with a myriad of potential artifacts that may complicate the identification of cognitive and psychosocial dysfunctions along with their causal mechanisms. What makes the study of SCA development so formidable? After all, in contrast to the more generic fields of learning disabilities and mild mental retardation, where the etiology is often unknown, SCA researchers begin their work by selecting individuals for study who have some discernible genetic abnormality. Should it not then be easier to link developmental behavioral sequelae with their known genetic origins?

A more thoughtful analysis soon yields the realization that chromosomal anomalies have complex and variegated effects on developing physiological and biochemical systems. Moreover, a range of environmental variables may either attenuate or potentiate the influence of biological determinants on behavioral development. Naturally, such factors are very difficult to disentangle. Furthermore, the SCA investigator encounters particularly complex problems associated with the choice of appropriate control groups, along with a host of other potential methodological pitfalls inherited from psychological allied domains. Perhaps, then, it becomes clearer why this field simultaneously offers a fascinating yet often frustrating endeavor for behavioral researchers.

The purpose of this chapter is threefold: to examine selected methodological problems encountered by SCA researchers, provide a critical appraisal of putative solutions, and to assess the applicability to SCA research of recent methodological advances culled from related fields of study.

In the first section of the chapter, a variety of matching procedures are described in order to facilitate the selection of appropriate comparison groups for use in SCA research. The subsequent and most extensive section is devoted to a detailed analysis of the assessment of cognitive dysfunctions from psychometric, information-processing, and neuropsychological perspectives. This is followed by a discussion of both practical and methodological difficulties that typically emerge during the course of longitudinal research. Finally, the topic of structural equation modeling is introduced because of its potential utility for devising and testing hypotheses regarding multivariate patterns of causal factors that no doubt influence the development of SCA individuals.

## MATCHING PROCEDURES

Most studies of SCA children include one or more comparison groups. Such groups are usually matched on factors that are judged as constituting potential confounding variables. For matched group designs, a given variable is generally considered a potential confound if it is at least moderately correlated with the dependent measure (a rough rule of thumb is a correlation of .40 to .70). Although matching procedures are necessary for many kinds of clinical research involving special populations, standard statistics and design textbooks in psychology and education offer only a modicum of information regarding matching techniques. Consequently, various alternative matching procedures will be described here in some detail, based in large part on a summary of the statistical literature provided by Anderson et al. (1980). As it turns out, most of the theoretical work in this area is applicable to SCA research, since it has been concerned with situations involving a dichotomous risk variable and a continuous confounding (matching) variable.

### Pair Matching

One of the two basic methods of forming matched groups is known as pair matching. With this type of procedure, one finds a specific comparison subject for each treatment subject. Pair matching techniques include caliper, nearest available, and stratified matching. Caliper matching consists of an attempt to form pairs of subjects who are equivalent by virtue of differing on the numerical confounding variable by no more than a small tolerance. For example, where chronological age (CA) is a potential confounding variable, one may decide to find matches who differ in age by no more than plus or minus six months. When the treatment subjects vary widely on such a variable, a larger tolerance may be permissible. Thus, in studies of selected samples of adolescents and young adults with Turner syndrome, to obtain a large enough sample (and these are usually considered small by most standards, e.g., 10 to 30 subjects), one often has to include individuals varying extensively in CA (Waber, 1979 [13 to 23 years]; Rovet & Netley, 1982 [Exp. I, 11 to 28 years]). In such cases investigators have adopted tolerances of up to three years.

Naturally, there is a tradeoff between size of tolerance and ease of finding comparison subjects. That is, the smaller the tolerance (desirable for reducing possible bias), the larger must be the pool of potential comparison subjects. As Anderson et al. (1980) point out, if the composition of the comparison reservoir is unknown, it becomes difficult to select a tolerance value that will produce a sufficient number of matched pairs. This could arise, for example, in a study of 45,X women to be matched on Verbal IQ from the Wechsler scales, where the potential comparison subjects consist of next-patient-controls who enter a hospital on an outpatient basis and present with symptoms leading to a diagnosis other than SCA. In this type of situation, one could not scan the values of potential comparison subjects beforehand. One possible negative consequence is that the tolerance value chosen could turn out to be too small to obtain matched controls for each subject. This would necessitate dropping the unmatchable 45,X women from the study, resulting in a smaller than adequate sample size. Under such conditions, one is advised to estimate the distribution of scores of potential comparison subjects by examining available institutional records that include the necessary information.

If the use of caliper matching with a relatively small tolerance is likely to result in some treatment subjects not being matched, an alternative pair matching method may be used: nearest available matching. By being less restrictive than the caliper matching procedure, this technique ensures that one can obtain a matched control for each treatment subject. It eventuates in a match by permitting the researcher to select from the remaining (not-yet-matched) individuals in the comparison pool the control subject who has the value (on the confounding variable) closest to that of a designated treatment subject. The sequence in which the treatment subjects are arranged is determined by one of three possible procedures: randomly ordering them on the values of the confounding variable; ranking them from the lowest to the highest values; or ranking them from highest to lowest values. With the nearest available matching approach, the comparison pool does not have to be larger than the treatment group. However, its effectiveness in reducing bias due to the confounding variable is limited to situations in which the relationship between this factor and the dependent variable is linear. The caliper procedure is effective in both linear and nonlinear cases.

A third pair matching method, stratified matching, is useful when working with a continuous confounding variable for which the scores are categorized into two or more strata. After establishing the categories, the researcher randomly forms individual matched pairs within each one.

## Nonpair Matching

When employing nonpair matching techniques, one does not select a specific comparison subject for each treatment subject. With the frequency matching method, for example, the distribution of the confounding variable is stratified for the treatment group. Then either additional comparison subjects are drawn from the pool (randomly or by stratified sampling) or some of those previously selected are dropped until there is an equal number of treatment and comparison subjects within each stratum. However, the number of subjects across strata may differ.

An alternative nonpair method is that of mean matching. Here one attempts to form equivalent distributions with regard to the confounding variable by equating the means of the treatment and comparison samples. A number of algorithms may be used to select members of the comparison group so that the difference between the means of the treatment and comparison groups is as small as possible. However, because the utility of mean matching depends strongly on satisfying the assumption of a linear response relationship, it is not a recommended procedure (Anderson et al., 1980).

## Multivariate Matching

The univariate caliper and stratified matching techniques described above can be adapted for use in multivariate matching. As the methods are relatively straightforward extensions of the univariate procedures, they will not be discussed further. Alternative multivariate techniques include: (1) minimum distance matching, a metric procedure in which two subjects can be matched if their scores on the confounding variables are close, as defined by any of a number of distance measures, and (2) linear discriminant matching, in which multiple regression analysis is used to determine the linear combination of the confounding variables yielding the best prediction of group membership. One of the univariate matching procedures described above is then applied to this single, linear discriminant function.

While one prominent study of intelligence in 45,X females has made use of multivariate matching (Garron [1977] matched his subjects on no less than eight variables: chronological age, race, years of schooling, residence [urban, suburban, rural], social class, marital status, ethnicity, and religious background), most investigations of SCA individuals have matched on no more than one or two variables. As Rovet (Chapter 3) points out, such studies are inadequate to the extent that they have not entirely accounted for the full range of possible confounding variables. Rovet suggests further that one way to circumvent the problems associated with selecting appropriately matched controls for SCA subjects is to study twins, one of whom has an abnormal karyotype and the other, a normal sex chromosomal complement. Despite the relative rarity of such conditions, Rovet has managed to find and study three such pairs in which one of each has a 45,X karyotype.

## Additional Considerations for SCA Research

In order to be able to validly attribute any behavioral differences between SCA and comparison groups to biologically-based differences (i.e., chromosomal constitution and ensuing biochemical and/or neurophysiological differences), investigators must make judicious decisions regarding the nature of the matching procedures they choose to employ. Certainly, the nature of the hypothesis being tested should determine to some extent the types of nonmatching characteristics of the comparison subjects that could potentially illuminate biological mechanisms of behavioral dysfunction. As an example, in attempting to determine if the neuropsychological impairment of 45,X women is lateralized, Pennington et al. (1985) incorporated not only normal (not brain damaged) controls (apparently mean-matched for CA and education level), but also three groups of brain-damaged controls (also matched, albeit loosely, for CA and education level). These women had either diffuse or localized (either left or right hemisphere) cerebral lesions as evidenced by an independently confirmed review of neurological data. While the 45,X women showed evidence of neuropsychological impairment relative to the normal females, their pattern of data was similar to that of diffuse-lesioned and normal females but different from that of either of the unilaterally-lesioned controls. The use of this design thus permitted the authors to conclude that the impairment of 45,X women does not appear to be lateralized.

## THE STUDY OF INTELLECTUAL, COGNITIVE, AND NEUROPSYCHOLOGICAL DYSFUNCTIONS

Despite a growing interest in the assessment of psychosocial problems associated with having a sex chromosome abnormality (see McCauley, Chapter 4), the domain of study receiving the greatest attention from SCA researchers has been that of intellectual abilities. Consequently, the bulk of the present chapter is concerned with many of the important methodological issues and problems that pertain specifically to the measurement and interpretation of cognitive dysfunctions. To this end, the first section consists of an analysis of statistical and normative factors that have a significant bearing on the validity of interpretations of discrepancies between measured Verbal and Performance IQs. The next section is devoted to a critical examination of both the conceptual and methodological problems involved in the ascertainment of differential deficits in cognitive functioning. Then the information-processing paradigm is described, along with subtle problems that frequently arise when using response times for studying mental processing. The role of microcomputers in future cognitive assessment is briefly examined in the subsequent section, followed by a relatively intensive discussion of numerous methodological considerations one should take into account when using noninvasive neuropsychological procedures for detecting lateral asymmetries in neurocognitive functioning.

## Verbal-Performance Discrepancies

Although most SCA investigators have used an extensive array of cognitive measures, scores on standardized intelligence tests continue to provide critical information for both clinical and research purposes. Moreover, since the Wechsler Intelligence Scale for Children-Revised (WISC-R) has emerged as the preferred instrument for use in the differential diagnosis of learning disabled children (Berk, 1982), it comes as no surprise that this instrument has also proven useful in SCA research and evaluation. Similarly, the Wechsler Adult Intelligence Scale (WAIS) has been used for many years in the study of SCA adolescents and adults, particularly with 45,X females. One of the major advantages of the Wechsler scales is that they provide separate IQ values for verbal and performance abilities. As such, one can compare these two scores to determine if there is any sizable discrepancy between them. Indeed, findings regarding various types of Verbal-Performance (VIQ-PIQ) discrepancies have constituted fundamental cognitive data for SCA researchers, often leading to follow-up tests designed to tap more specific processing deficiencies underlying the lower, more generic Verbal or Performance IQ. For example, early studies with the WISC and WAIS revealed that while 45,X females exhibited average to above average Verbal IQs, their Performance IQs tended to be relatively depressed (Money & Alexander, 1966; Shaffer, 1962).

The major rationale for covering this topic in the present chapter has to do with the manner in which such discrepancies should be interpreted. As it turns out, there are a number of both scientific and statistical complexities associated with this endeavor that must be understood to avoid misinterpreting the psychological importance of given magnitudes and/or directions of Verbal-Performance discrepancies. Drawing on the ideas of Silverstein (1981), Berk (1982) has provided a useful framework for evaluating discrepancy scores derived from the WISC-R. Elucidating the meaning of these scores requires a separate consideration of the reliability, abnormality, and validity of Verbal-Performance discrepancies. For a given discrepancy score to be reliable, it must be so large as to be unlikely to have occurred solely by chance or errors of measurement. As computed by Wechsler (1974, p. 35), 12 points constitutes a statistically significant discrepancy at the .05 level of confidence, and 15 points at the .01 level. Regardless of their direction, discrepancies that equal or exceed these values may be considered reliable.

While a reliable discrepancy is necessary for determining whether an abnormality exists, it is not sufficient for making such an inference. Rather, as Berk (1982) describes it, an abnormal discrepancy is one so large that it was obtained by only a relatively small proportion of the children in the WISC-R standardization sample. This so-called normative approach was developed by Kaufman (1976), who reported the frequency with which discrepancies of various magnitudes occurred within the normal population, as indexed by the incidence rates for the 2200 children in the standardization sample. These values can serve as base rates for determining abnormal discrepancies. Surprisingly, close to 25% of the normal population have a discrepancy of at least 15 points, a value that Wechsler (1974) suggested was important and worthy of further investigation. While a score of 12 points (significant at the .05 level of confidence) occurs in as many as 34% of the normal population, larger discrepancies occur much less frequently (e.g., 20, 25, and 30 points occur at the rates of 12%, 4% and 2%, respectively). The interested reader should consult Sattler (1982, p. 564) for a more detailed table that provides values for 11 age groups ranging from 6.5 to 11.5 years. Normative rates are also available for the Wechsler Preschool and Primary Scale of Intelligence (WPPSI) (Sattler, 1982), the WAIS (Field, 1960), and the WAIS-R (Grossman, 1983a). Furthermore, Berk (1982) provides a brief review and critique of formulas that have been developed for computing the abnormality of a Verbal-Performance discrepancy.

The user of the WISC-R normative tables should be cautioned at this juncture about a potential problem concerning the percentage values. Grossman (1983b) has pointed out that the frequency data provided for given magnitudes reflect discrepancies irrespective of their direction. Therefore, one cannot accurately obtain an appropriate base rate for a discrepancy of a particular direction (e.g., VIQ>PIQ) by reading it directly from published tables. However, since Kaufman (1976) found that approximately equal numbers of children from the standardization sample exhibited VIQ>PIQ as PIQ>VIQ discrepancies, Grossman suggests that one can simply halve the tabled values to yield appropriate incidence rates. At this point it should be noted as well that determining the existence of an abnormal discrepancy provides no additional information regarding the underlying nature of the abnormality. As Kaufman (1979) describes it, if half a dozen children produce discrepancy scores of 15 points in the same direction, there may nevertheless be entirely different bases for their common performance patterns.

The third factor considered by Berk (1982) is the validity of reliable discrepancy scores. Specifically, this refers to the accuracy of such scores in differentiating children who have learning disabilities (as assessed through other means) from those who do not. Reviewing the extant literature in which magnitudes of discrepancy scores are compared, Berk concludes that the findings are equivocal at best and calls for further research to clarify the issue.

The interpretive problems described above have important implications for SCA researchers, since obtained Verbal-Performance discrepancies have historically provided the initial impetus for exploring more detailed characteristics of verbal and performance deficits. For example, a number of studies have documented consistently larger discrepancy scores for certain types of SCA individuals as compared to matched control groups (see Chapter 3 for a review of such findings with 45,X females). However, closer inspection of these data reveals that the magnitude of the average discrepancy score is sometimes too low to be considered reliable (e.g., Garron, 1977), and even when high enough to be reliable, may still not be large enough to be interpreted as abnormal when compared to frequencies from the appropriate standardization sample. Moreover, even if the average discrepancy score obtained from a given study is reliable and relatively large, there may also be a large degree of variability within the SCA sample. Rovet (Chapter 3) notes such variability in a sample of 45,X women as reported by Pennington et al. (1985). Similarly, Berch and Kirkendall (1986) have found a wide range of discrepancy scores in 45,X children, with half exhibiting comparatively large and reliable, positive discrepancy scores and half showing relatively small, nonsignificant scores (some positive, some negative). Such a sizable lack of consistency within an SCA subtype obviously complicates the interpretation of a significant average discrepancy as reflective of some pervasive cognitive dysfunction.

Finally, the meaning and importance of exactly the same discrepancy score may vary depending upon the absolute Verbal and Performance IQ scores. For example, suppose two children have a 20-point discrepancy score, one of whom has a PIQ of 120 and a VIQ of 100, and the other, a PIQ of 100 and a VIQ of 80. Certainly, the clinical interpretation of these discrepancies would differ for these two children. While this is quite obvious to most clinicians, SCA researchers for the most part have not as yet fully examined the theoretical implications of such distinctions. For example, this author has tested a young 45,X girl who exhibited a VIQ-PIQ difference of 33 points, constituting both a statistically and normatively large discrepancy. Yet her VIQ and PIQ scores were 133 and 100, respectively. While her pattern of scores replicates those of other children and adults with Turner syndrome, she clearly does not have inferior spatial abilities (at least as indexed by the Performance IQ). A thorough understanding of the cognitive processing operations of SCA individuals will require a more thoughtful interpretation of such scores than has previously been provided.

## Differential Deficits in Cognitive Processing

In order to interpret the results of behavioral assessments of SCA individuals, investigators must compare their scores with those of unaffected individuals. In many instances, the performance of SCA subjects will be poorer. How should one interpret such a finding? A survey of the broader literature in the field of developmental disabilities reveals that researchers often consider poorer performance to be reflective of one or more of the following: deficit, defect, dysfunction, deficiency, difference, disability, disorder, or impairment. The choice of a given term can be based on a particular meaning that the investigator attributes to it. More often than not, however, a researcher will use some of these terms interchangeably, without specifying the intended meaning(s). While dictionary definitions help, they are no substitute for more precise, theory-based definitions that can be linked to explicit empirical referents. For example, the terms deficit and deficiency both have lexical meanings of "incompleteness" or "lacking something essential," while the term "defect" refers to an "imperfection" or something that is "faulty."

What does it mean to say that an individual has a "cognitive deficit"? Should this imply that one suffers from a permanent defect? Stanovich (1978), for one, suggests that the term deficit carries such a connotation. Does having a permanent defect necessarily mean that the associated cognitive dysfunctions are not remediable? How does one establish that a "difference" in performance on a given behavioral measure reflects a "differential deficit"? Indeed, the latter problem has received considerable attention in the mental retardation literature. Since most of the methodological issues treated in this literature are directly relevant to the design and interpretation of SCA research, they will be discussed here in some detail.

The use of single tests. One major problem involves the interpretation of a performance difference between two groups of subjects on a single test. Suppose that an SCA group performs more poorly than a control group on the Visual Sequential Memory subtest of the Illinois Test of Psycholinguistic Abilities (ITPA). We would be wrong in concluding that the SCA group has a deficit in visual sequential memory. According to Torgesen (1975), such a conclusion is an example of committing the so-called "stimulus error." This refers to making the mistake of identifying test responses and the processes they purportedly measure with the name of the test. This particular ITPA subtest was chosen as an example because a process-oriented study has revealed that successful performance on this putatively nonverbal, visual task is largely attributable to the use of a verbal labeling strategy (Bowen, Gelabert, & Torgesen, 1978).

Furthermore, issues of construct validity arise when poor scores on a given test are interpreted as reflecting hypothetical attributes of the examinees. Stanovich (1978) similarly notes that "the observation of a performance difference on a single task sheds no light on which particular processes might be implicated" (p. 32). Milgram (1973) contends that attributing a deficit on such a basis is an instance of confusing correlation with causation. In addition, Elliot (1970) suggests that performance differences may be attributable to nonspecific factors such as motivation, incentive, and attentiveness. As Torgesen (1979) cogently argues, in order to adequately assess the psychological processes underlying responses on a given test, one must have a well-developed theory that designates the processing activities required for successful performance. Then if an individual scores relatively low on the test, follow-up tests that vary the task's major parameters can be used for identifying the basic processing difficulty responsible for the poor performance.

For example, suppose that as a group, SCA children score comparatively poorly on the Digit Span subtest of the WISC-R. We cannot conclude safely from this test performance alone that SCA children have a deficit in short-term memory or that they have sequencing problems. Their poorer performance could be attributable to a smaller memory capacity, a relatively inefficient use of memory strategies such as chunking and rehearsal, or the speed with which they initially encode (i.e., identify) the numerical stimuli. Isolating the underlying mechanism would require the manipulation of critical task parameters and subsequent comparison of the ensuing performance patterns. As it turns out, the bulk of the evidence from experimental studies suggests that speed of encoding is the primary determinant of developmental increases in digit span (Dempster, 1981). While this same factor does not appear to explain individual differences in size of digit span for intellectually normal adults, Das (1985) has provided evidence that it may be responsible for the poorer digit span of mildly retarded preadolescents. Finally, Berch, Hartmann, & Bofinger (1986) recently reported that the poorer digit span performance commonly exhibited by adolescent and adult women with Turner syndrome may also be attributable to a slower speed of encoding.

Psychometricians have attempted to identify more basic psychological processes underlying performance on specific subtests through the use of factor analysis. However, this approach may only compound the problems. For example, Kaufman (1975) conducted a factor analysis of the WISC-R that yielded three factors. While the first two generally reflected the verbal and performance dimensions built into the test, a third factor emerged from a loading of the Arithmetic, Coding, and Digit Span subtests. Following Cohen's (1957, 1959) analysis of the WAIS and WISC, this has been labeled the Freedom from Distractibility factor. Should one conclude, then, that examinees who perform poorly on this combination of subtests are more "distractible" than those who score higher?

A relatively recent, process-oriented study appears to refute such a conclusion. Stewart and Moely (1983) examined relationships between performance on the three relevant subtests and a variety of cognitive and behavioral measures for 96 fifth graders. The battery of cognitive measures included an extensive array of laboratory (process-based) and psychometric measures, while the behavioral indices included the recording of looking, naming, gesturing, and distractive (off-task) responses exhibited during one of the recall tasks. The results indicated that performance on the Arithmetic subtest may be reflective of not only numerical ability, but also the use of a rehearsal strategy (to recall the information provided in the task), an ability to comprehend and integrate verbal information presented in a mathematical context, and an awareness of when the appropriate response has been made. Performing well on the Coding subtest appeared to require an ability to deal with numbers as well as the conscious use of a rehearsal strategy. In contrast to the Arithmetic and Coding subtests, only performance on a visual memory span task (with pictures of common objects) was related to performance on the Digit Span subtest. Moreover, distraction (i.e., attending to incidental aspects of the environment or looking around aimlessly) almost never occurred during the task in which it was being measured. Stewart and Moely point out that this result may be attributable to the highly structured nature of the particular task itself. Yet they note further that the three WISC-R subtests are also highly structured. They conclude that distraction may not be the key factor responsible for inhibiting performance on these subtests. Rather, performance levels are apparently attributable to differing and fairly complex cognitive processes.

   The ability-by-treatment interaction approach. In applying a process orientation to the study of differential deficits, investigators have employed what has become known as the ability-by-treatment interaction approach. The rationale seems fairly straightforward. As Detterman (1979) describes it in relation to mentally retarded populations, in order to demonstrate a differential deficit, one must first show that both mentally retarded and normal subjects perform similarly under some level of an independent variable. This should rule out the possibility that the ability being studied was confounded with some other variable. If a group difference were to be found at another level of the independent variable (i.e., a statistical interaction), one could conclude that this difference reflected a deficit in the ability being studied. However, as Detterman points out, even if the ideal interaction is obtained, several artifacts may exist that could invalidate the conclusion of a differential deficit.

   The most pernicious of these are floor and ceiling effects. If a task is so difficult that all subjects perform at chance levels (floor effect) or so easy that performance is nearly perfect for all subjects (ceiling effect), existing group differences could be concealed, rendering a resulting statistical interaction as uninterpretable (Baumeister, 1967). Thus, as Detterman (1979) describes it, for one to achieve an unambiguous interpretation from any study comparing mentally retarded (or any clinical population) and normal subjects, floor and ceiling effects must be ruled out, and both groups must perform equivalently for at least one level of the independent variable. However, Detterman makes the following rather startling observation: "In the hundreds of studies that have been conducted comparing normal and retarded performance, few, if any, have yielded results that satisfy these criteria even when substantial efforts were directed toward their satisfaction" (p. 729). Yet he goes on to note that while most investigators would insist upon the elimination of floor and ceiling effects, they are less stringent about the equivalent performance criterion. That is, comparisons yielding a larger performance difference at one level of an independent variable than at another would be interpreted as a relative deficit. Unfortunately, as other researchers have pointed out, the interpretations of such noncrossover statistical interactions are tenuous at best (Bogartz, 1976; Loftus, 1978; Pachella, 1974; Stanovich, 1978). Briefly, this problem arises because most dependent variables in psychology (e.g., percent correct responses) are merely arbitrarily related to the underlying constructs that they purportedly measure. Consequently, noncrossover interactions can be eliminated or created by rescaling the dependent scores via several types of monotonic transformations, no one of which may be more theoretically justifiable than any other.

As if these problems were not enough to deter even the most dedicated investigator, more subtle but no less pernicious methodological problems await those interested in assessing differential deficits. Chief among these are item difficulty and task reliability. Suppose that one is attempting to compare the performance of SCA and chromosomally normal subjects on two tasks (which comprise different levels of a given independent variable), wherein the reliability of one task is greater than the other. Chapman and Chapman (1973, 1974) have pointed out that other things being equal, the more reliable task will yield a larger mean difference between the groups being compared. Furthermore, reliability is in part a function of task difficulty, with tasks of moderate difficulty being more reliable than either relatively easy or relatively difficult tasks. In other words, tasks of intermediate difficulty have the greatest discriminating power (i.e., the extent to which a given test differentiates two groups in the ability measured by the test). As Detterman (1979) points out, this means that the finding of a differential deficit in any given study comparing clinical and normal populations could as readily be a function of the levels of task difficulty employed as of any real deficit (see Chapman & Chapman, 1974, for detailed examples of this type of problem). Suffice it to say that Chapman and Chapman (1974) consider this problem serious enough to suggest that the ability-by-treatment study design "yields artifactual findings so readily that one must question the findings of most studies in which it has been used" (p. 404).

How can one resolve this problem? Chapman and Chapman (1974) offer a solution that involves comparing subject groups on two levels of two tasks rather than two levels of a single task. For the normal subjects, these tasks must be matched at each of the two levels of the independent variable for mean level of performance, variance of item difficulty, shape of distribution of item difficulty, and reliability. If a larger ability-by-treatment interaction is obtained for one task as compared to the other, the clinical subjects can be concluded to be deficient on that task relative to the other. According to Detterman (1979), while this solution is comparatively straightforward in principle, it would be very difficult to accomplish in practice. Moreover, even though this approach is methodologically satisfactory, the kind of interpretation it yields cannot be expressed in any absolute sense, and thus is difficult to translate into psychologically meaningful terminology (Detterman, 1979).

This rather dismal state of affairs seems to suggest that almost any study of differential deficits is likely to contain at least one methodological flaw that renders its findings potentially artifactual. Following Detterman (1979), any conclusions drawn from such studies must be viewed as extremely tentative. He concludes that, "We unfortunately, are making the best of a very bad situation" (p. 732). It is also noteworthy that Detterman has since carried out a study with mentally retarded and nonretarded adults in which he went to great lengths to take into consideration most of the problems described above when analyzing and interpreting his data (Caruso & Detterman, 1983). In this experiment, mildly retarded young adults were found to make more errors and respond more slowly than nonretarded controls on a laboratory task of stimulus encoding (matching a 16-cell, checkerboard-type matrix of filled and empty squares to a target stimulus aligned in a linear array of three other, nonmatching matrices [i.e., similar to a Wechsler Block Design pattern]). However, by having assessed performance patterns across a wide range of stimuli and a representative range of stimulus difficulty, Detterman was able to carry out correlational analyses that revealed a high degree of similarity for the normal and retarded groups. This suggests that mentally retarded adults encode stimuli in much the same way as nonretarded individuals, albeit at slower rates (i.e., they do not suffer from a differential deficit).

The implications for comparative research in the area of developmental disabilities in general and sex chromosome abnormalities in particular are clear. We must interpret our data with extreme caution until more satisfactory methods are developed for resolving the methodological problems that are inherent in research aimed at discerning dysfunctional psychological abilities in clinical populations.

### Information Processing Approaches

Among the various information-processing procedures that have potentially important utility for the study of cognitive dysfunctions, so-called chronometric or response-time techniques constitute perhaps the most significant ones for SCA researchers. Thus far, the primary applications of such techniques have been carried out with 45,X females (Berch, Hartmann, & Bofinger, 1986; Berch, Kirkendall, Briscoe, Dignan, & Smith, 1985; Rovet & Netley, 1982). The use of response-time procedures for the assessment of cognitive processing dysfunctions in SCA individuals is likely to increase rapidly as their benefits become more widely recognized and packaged software for microcomputers becomes more readily available. However, the danger is that some of the complexities inherent in the use of these techniques will not be as easily accessed from the somewhat esoteric literature of experimental cognitive psychology. Consequently, I will provide the reader with a relatively detailed tutorial covering the conceptual bases for the use of response-time procedures, along with specific methodological problems that arise in their application to the study of cognitive development. In addition, I will discuss issues pertaining to the statistical analysis of response-time data and to other problems that emerge when interpreting the meaning of slower response rates obtained from clinical populations.

Stage analysis. The standard response-time technique involves the recording of latencies, usually in milliseconds, measured from the onset of the stimulus to the onset of the response. The stage analysis of response times includes two interrelated enterprises: decomposition of response times into component, functional units called stages, and analysis of processing within such stages (Taylor, 1976). Depending upon the specific task, these may include encoding, transformation, comparison, decision, response selection, and response execution. Encoding itself may be composed of substages that include stimulus preprocessing, feature extraction, and item identification (Sanders, 1980); the transformation stage may consist of rotating a mental image of an object to an upright orientation; the comparison operation might characterize the process of determining whether a retrieved mental representation of a stimulus item is equivalent to a stimulus probe, the decision stage could involve a judgment as to whether an internal representation matches an external probe; response selection consists of choosing the appropriate effector muscles for generating a motoric response (e.g., pressing one of two response buttons or vocalizing one of two words); response execution refers to the actual movement involved in making a response.

Much of the modern application of response-time techniques is derived from the additive factors method developed by Sternberg (1969). As outlined by Chase (1978), Mulder (1983), and others, the additive factors method is based on four assumptions: information processing is organized into a sequence of independent stages; each stage operates on an input from the preceding stage; each stage operates on its input independently of any other stages; the total response time is the sum of the processing times for each separate stage. Furthermore, as Chase has pointed out, the notion of "processing" independence proffered here should not be confused with that of "statistical" independence. The latter involves a much stronger constraint, requiring that there be no correlation between durations of different stages. Processing independence can occur without statistical independence (which may be due, for example, to a general state of arousal that yields relatively large correlations between component times for the various stages). While the mean component response times would still be additive under such conditions, the variances would not be.

Sternberg's additive factors technique involves the manipulation of multiple task variables in an effort to determine the consequent pattern of effects. The logic of this technique dictates that if two (or more) task variables influence different processing stages, they will produce independent, additive effects. In contrast, if these variables influence the same processing stage, they should interact (in a statistical sense).

As Chase (1978) and Mulder (1983) point out, the additive factors method has a number of limitations. First, it is not powerful enough to reveal durations of separate stages. For this, one needs to employ the "subtractive" technique in conjunction with the additive factors method. This approach, based on the original work of Donders (1868/1969) and elaborated upon by Sternberg (1966), consists of administering a series of tasks, which through successive deletions enables the researcher to estimate the amount of time taken to execute a component mental operation.

Another limitation is that the additive factors method yields no direct indication of the order of stages. Furthermore, it does not provide a direct suggestion as to the substantive interpretation of the various stages. Finally, if an experimental variable influences the output of a stage as well as its duration, the effect of one stage may be carried over to a subsequent stage. This state of affairs would make quite tenuous the basic assumption concerning the means by which separate stages are identified. In spite of the limitations described above, the additive factors method has proven most useful as an initial step in the study of information-processing sequences by establishing the existence of stages and the variables that influence them.

Response-time analysis of cognitive development. Kail and his colleagues (Kail, 1985; Kail & Bisanz, 1982) have discussed some of the important methodological issues associated with the use of response-time methods for studying cognitive development. One major issue is that of the reliability of response-time data. Since these techniques are typically used for detecting very small task differences, frequently on the order of 250 milliseconds or less, measurement error has to be minimized. This is usually accomplished by the following procedures: (1) adopting a repeated-measures or within-subjects experimental design, in which every subject encounters all conditions of the task; (2) administering a relatively large number of "warm-up" trials, permitting the subject to become familiar with the task and to select a single processing strategy; and (3) administering a large number of test trials and then analyzing the mean or median response times for the various conditions.

Kail and Bisanz (1982) point out a potentially problematic side effect of the procedures just described. They suggest that each of these may increase the amount of practice a child receives. Furthermore, this side effect can yield various complications for interpreting response-time data (which the authors suggest are frequently ignored): extended practice may modify cognitive processing significantly; if practice produces systematic changes over trials, and if the order in which the different experimental conditions (in a repeated-measures design) is not counterbalanced or randomized across subjects, then an apparent difference in cognitive processing could actually be attributable to practice per se; since extended practice may produce cognitive processing quite different from that used in settings other than an experimental one, the generalizability of any conclusions may be lessened considerably.

In addition, Kail (1985) has pointed out another problem concerning the complexity of comparing the results of studies differing in the number of "warm-up" trials. In this regard, he discusses two developmental studies of memory scanning, one in which 240 practice trials were administered and another with only 16 trials. As Kail notes, it is not surprising that the subsequent test trials from these studies yielded very different estimates of processing speed. The interested reader should consult Kail (1985) for some procedures that may be used to remedy this situation.

Statistical issues in response-time analysis. Another major methodological issue concerns the statistical analysis of response-time data. Pachella (1974) points out that the mental operations studied by cognitive psychologists are considered to fill real time. Since response time is used to measure the real-time duration of these mental events, it constitutes a ratio measure, and therefore, does not require a monotonic transformation of scale (e.g., logarithmic or reciprocal) to make the data suitable for statistical analysis. As Stanovich (1978) notes, one cannot overemphasize the importance of a dependent measure such as response time that is immune to arbitrary rescaling. This is indeed of critical importance for the study of differential deficits via use of the ability-by-treatment interaction design. That is, since arbitrary rescaling of response-time data is unwarranted, the occurrence of statistical interactions, even noncrossover ones (the most frequently occurring type), can be interpreted in a comparatively straightforward manner.

Speed-accuracy tradeoff. Other problems may arise with regard to the error data in chronometric tasks. Stanovich (1978) states that one of the most overlooked problems is that the variability in response times may be partially attributable to variability in error rate. This can occur because of the so-called speed-accuracy tradeoff. Specifically, since subjects are not usually instructed explicitly as to a desired level of accuracy or speed of response, they must set their own criterion to a response speed that will yield an acceptably high level of accuracy. Typically, the faster that one responds the higher the error rate. If subjects' speed-accuracy criteria are correlated with experimental conditions (or subject groups, such as intellectually normal and mentally retarded), interpretation of the response-time outcomes becomes problematic (for a more complete discussion, see Pachella, 1974). Stanovich suggests that researchers usually attempt to handle this problem either by trying to obtain error rates that are invariant across experimental conditions or by performing an analysis of covariance on the response-time data using error rate as a covariate. Other approaches are discussed by Kail and Bisanz (1982). Stanovich also emphasizes that keeping the overall error rate low does not constitute a satisfactory precaution. Along similar lines, Kail (1985) illustrates how even a seemingly negligible difference in error rates can complicate the interpretation of response-time parameters. In any event, Stanovich argues that investigators should, at the very least, always report error-rate data, including both overall rates and rates for the various experimental conditions.

It should also be noted that an additional problem can arise when comparing the response-time performance of special populations and normal subject groups. Namely, if a clinical group exhibits a slower response rate than a normal group, this does not necessarily mean that the clinical group adopted a more lenient speed criterion (i.e., that they attempted to respond too slowly). Rather, it is possible that despite their comparatively slow response times, they may still have been responding too quickly (for them) to achieve a higher accuracy rate (Brunner, Berch, & Berry, 1987).

Kail and Bisanz (1982) describe some other methodological concerns. Among these are (1) as stimulus complexity and the number of alternative responses available to the subject increases, response-time methods and analyses both tend to become rather unwieldy; and (2) difficulties arise in determining if a given information-processing model "sufficiently" explains the data, as well as in comparing the relative superiority of alternative models. Kail and Bisanz conclude, nevertheless, that response-time methods can be a valuable tool for revealing mental processing when used appropriately.

## Computerization of Cognitive Assessment

What does the future portend with regard to the nature of assessment devices that will measure the types of intellectual processing that are likely to be of interest to SCA researchers? Along with other as yet unknown developments, microcomputers will no doubt play a significant role in the design and standardization of mental tests as well as in the administration of both new and old instruments. A recent critical appraisal of computerized testing possibilities led Hunt and Pellegrino (1985) to conclude that while computerized presentation for testing verbal comprehension will yield relatively little difference from conventional procedures, it could extend the ways in which we currently evaluate spatial-visual reasoning and memory. They also state that computer-controlled item presentation may lead to the development of tests of learning potential and other psychological functions that are not generally evaluated by conventional intelligence or aptitude tests. Recently, a report has appeared in the literature describing a new battery of microcomputer-based spatial tests. The software package has been made available for a nominal fee (Pellegrino, Hunt, Abate, & Farr, 1987).

## Neurocognitive Assessment

As is evident in the chapters by Rovet and Netley in this volume, neuropsychological methods are being used increasingly for studying SCA individuals. Naturally, space limitations preclude any discussion of the countless extant measures that may be employed by the investigator whose purpose is to assess focal central nervous system dysfunctions. For descriptions of clinical neuropsychological batteries along with a host of other paper-and-pencil tests, the interested reader should consult other sources (see Chadwick & Rutter, 1983; Feuerstein, Ward, & LeBaron, 1979, for tests used with children).

Techniques for studying the hemispheric organization of various cognitive processing operations may be particularly interesting to SCA researchers. These methods have begun and probably will continue to provide important information concerning how abnormal brain lateralization mediates behavioral expressions of X aneuploidy. The two most extensively used noninvasive methods in this regard are the dichotic listening and visual half-field paradigms. Consequently, a discussion of each will follow, aimed at providing the reader with: a brief account of the neuroanatomical pathways presumed to mediate the transmission of auditory and visual sensory information to the cerebral hemispheres; a description of the basic experimental procedures of each paradigm; a discussion of some of the major methodological issues associated with each of the approaches; and an evaluation of the applicability of such techniques to the study of clinical populations in general and SCA individuals in particular. Finally, some recent advances in neurological techniques such as brain imaging will be discussed briefly.

The dichotic listening procedure. Kimura (1961) adopted Broadbent's (1954) dichotic listening procedure in order to study auditory asymmetries in hemispheric processing. Presenting subjects with three successive pairs of digits (one of each pair to each ear), she found better report for those digits presented to the right ear. As outlined by Bradshaw, Burden and Nettleton (1986), she explained this outcome by making the following assumptions: one cerebral hemisphere (typically the left) is specialized for speech; auditory inputs are represented more strongly in the contralateral than ipsilateral hemisphere; since contralateral input suppresses information on ipsilateral pathways (partially if not completely), dichotic stimulation initially lateralizes the input; information from the nondominant ear (hemisphere) is transferred for processing across the commissures to the dominant (for speech) hemisphere, where it comes into contact and purportedly competes with the direct contralateral input from the right ear.

In evaluating these assumptions (see Bradshaw et al., 1986 for relevant references), Bradshaw and his colleagues suggest that ear differences may be indicative of either the relative processing superiority of one hemisphere or the loss of information during interhemispheric transmission to the only hemisphere that can execute the task. The second assumption has received support from studies demonstrating that the contralateral auditory pathway by-and-large yields more cortical activity and has more fibers. In contrast, evidence has recently been reported that is inconsistent with Kimura's assumption concerning ipsilateral suppression. Finally, despite earlier evidence from the study of commissurotomized patients supporting the fourth assumption, more recent findings contraindicate it.

Bradshaw et al. offer a number of criticisms regarding the significance of the finding of a dichotic right-ear-advantage (REA). Among these are that REAs are poor predictors of language lateralization, yield low intertest reliability, and their size and consistency often increase with extensive practice (not simply due to a reduction in response variability). They suggest further that while REAs may be a correlate of language laterality, they make poor indices of it. However, these authors admit that REAs measure receptive components rather than the more strongly lateralized expressive aspects that are clinically assessed. A review of research with a range of auditory stimulus materials led Bradshaw et al. to conclude that a single dichotic test (either verbal or nonverbal) is likely to underestimate the true incidence of language lateralization. Furthermore, they note that while the use of two tests representing the opposite extremes (verbal and nonverbal) may accurately index the direction of language lateralization, they cannot reveal the degree. In discussing the complexities associated with aligning the timing of stimulus inputs to the right and left ears, Bradshaw et al. suggest that in order to cancel any unwanted asynchronies, the same stimulus material should be replayed to the subjects a second time, reversing presentation to the two ears.

Bradshaw et al. conclude, nevertheless, that with proper safeguards, dichotic techniques can be usefully employed for clinical purposes. In addition, they suggest that the overall laterality effect may depend upon capacity limitations, the mix of required component processes, and the magnitude of interhemispheric transfer that may contribute noise to the transferred information. Consequently, they caution investigators to be careful in drawing conclusions regarding the degree or direction of lateralization of a total task based on the size of ear differences. With regard to the use of the dichotic listening procedure for studying lateral asymmetries in SCA individuals, the only peculiar methodological problem that might arise is with 45,X females. Namely, one must attempt to control for the relatively frequent degree of peripheral hearing impairments associated with Turner syndrome.

The visual half-field technique. The other major noninvasive approach used for evaluating lateralization is the so-called visual half-field or divided visual field paradigm. Each visual field projects initially to the contralateral hemisphere. As Moscovitch (1986) describes it, differences in perceptibility between stimuli confined to either the left or right visual field occur because stimuli in one field possess privileged access to the contralateral hemisphere specialized for processing them. Consequently, verbal stimuli (e.g., words) will typically be perceived both faster and more accurately when presented in the right visual field (RVF), while nonverbal stimuli (e.g., faces or nonsense figures) will triumph in the left visual field (LVF).

As outlined by Moscovitch (1986), the typical procedure involves stimulus presentation via a tachistoscope, with the subject's eyes initially directed to a central fixation point. Following a ready signal, a stimulus appears in the periphery for a duration of 200 milliseconds or less. This exposure time is used to prevent eye movements from visually centralizing the stimulus, which would produce the consequence of projection to both hemispheres. The stimuli may be presented unilaterally or two at a time, with one on each side of the fixation point. The subject's task is to either identify, recognize, categorize, or match the stimuli to each other or to a previously presented stimulus, with either accuracy or latency of the responses serving as the dependent variable. Stimulus materials have included words, faces, digits, line drawings and photographs of objects, pictures of scenes, dots, simple lines, and nonsense figures. According to Moscovitch, while no optimal strategy has yet been determined for presenting nonverbal material, words are best presented in a vertical orientation; this reduces the likelihood of directional scanning that can confound the interpretation of findings regarding hemispheric lateralization.

One of the critical problems for the investigator using this procedure is to be certain that the stimulus is restricted to the intended visual field. Moscovitch notes that while there is strong evidence that almost all normal subjects fixate their eyes accurately prior to stimulus presentation, this may not hold true for clinical populations. As a precaution, he suggests that catch trials be inserted in which a very small digit is presented centrally, so that it can be identified only if the subject is fixating appropriately. Subjects should be excluded from the experiment if they fail to identify this digit on more than 10% of the catch trials.

Moscovitch (1986) has leveled a noteworthy criticism at investigators who have used the visual laterality paradigm. He argues that they have been "cavalier" in describing their stimulus parameters as compared with those who use the dichotic listening paradigm. Specifically, he notes that while auditory researchers commonly report stimulus intensity, frequency, duration, and signal-to-noise ratios, studies of visual laterality rarely provide information about anything other than stimulus duration and eccentricity. Moscovitch indicates that such omissions are regrettable, in that recent research has demonstrated that factors such as luminance and contrast may have different effects on the left and right visual fields.

With regard to the study of individual differences in visual field asymmetries (including clinical populations), Moscovitch (1986) has offered some criticisms that have relevance for SCA researchers. He suggests that interpretations of such studies are derived from the assumption that visual field differences reveal either the extent of intrinsic hemispheric specialization or the size and integrity of interhemispheric pathways. He argues that in numerous instances this assumption, and the interpretations based on it, are either "ill-founded or inadequate" (p. 107). This is apparently due to difficulties involved in determining the extent to which visual field asymmetries can be influenced by variability in the level of hemispheric arousal or the adoption of strategies that are not under the control of the investigator. In addition, Moscovitch suggests that if hemispheric specialization is influenced by multiple factors, then laterality tests (even within a single modality) may not be assessing a unitary aspect of hemispheric specialization.

General methodological concerns. One of the prime issues concerning the use of any noninvasive lateralization measure is that of test-retest reliability. As Segalowitz (1986) points out, low reliability may stem from strategic or attentional shifts with practice as well as from inconsistencies in what the tests measure. After reviewing the relevant evidence, Segalowitz concludes that the most popular lateralization techniques have exhibited relatively high stability. Nevertheless, he notes that reliability levels for the visual half-field procedure are considerably lower than those obtained for the dichotic listening technique.

In addition to the issues described above, other complicating factors may arise for those interested in studying the developmental course of cerebral lateralization (Hahn, 1987). However, Bryden and Saxby (1986) suggest that good techniques are currently available for studying hemispheric specialization in children, and that there is little evidence indicating any systematic changes in cerebral lateralization after the age of 2 or 3 years. This information is of particular importance for researchers involved in testing theories of neuropsychological development in SCA individuals, such as that put forth by Netley (Chapter 6).

At this juncture, the reader may conclude that the list of potential complexities associated with the use of laterality techniques has been exhausted. Unfortunately, this is not the case. To summarize the variety of problems not yet discussed here, a passage is presented from a detailed monograph concerned exclusively with the study of human cerebral asymmetry (Bradshaw & Nettleton, 1983). While the statements were originally made with regard to the lack of replicable findings when these techniques have been used with normal subjects, most of the concerns are relevant for the testing of SCA individuals.

One reason for the often contradictory findings in laterality research with normal subjects is the great diversity of experimental techniques available for use. These may involve reaction time or accuracy measures; a speed or an accuracy set imposed on the subject; vocal or manual responding; go-no-go or target-nontarget responding; identity, similarity, or category (e.g., male-female) matches or discriminations; perceptual matching between two simultaneously presented stimuli versus matching a test item to a previously memorized target (with long or short retention intervals); constant or single target versus continuously changing or multiple targets. Moreover, the task material itself may be familiar or unfamiliar, easy or difficult, practiced or unpracticed, accompanied by another easy or difficult secondary task, of a verbal or nonverbal nature, or entirely on its own. It may be easily verbalized, subject to verbal recoding, or impossible to verbalize. It may be treatable as a unitary integrated configuration, or it may be loosely organized as a collection of isolated features or elements. These factors do not constitute a comprehensive list, nor probably are they independent of each other in their effects, and all may well affect the replicability of laterality studies. Even the employment of extrafoveal tachistoscopic exposures of brief duration (to ensure fixation) may *itself* affect laterality patterns (p. 84).

Certainly, selection of specific task parameters when using laterality techniques with SCA subjects is likely to constitute in itself a formidable task for researchers. Judicious decisions may be made best when primed by an appropriate combination of theoretical, methodological, and practical concerns. In any event, a detailed yet lucid exposition of the procedural characteristics in the method section of published articles should go a long way toward clarifying any conflicting findings obtained from different studies of purportedly the same phenomenon. Hopefully, both the differences and similarities in brain function characterizing SCA individuals as compared with chromosomally normal persons will be robust enough to be replicable across studies in which "slightly" different variations in task parameters may be employed. This remains to be seen.

Finally, it should be noted that extant EEG methods as well as recent developments in brain imaging techniques may find future application to the study of subtle neuropsychological dysfunctions in SCA individuals. Rovet (Chapter 3) reviews some studies of EEG wave patterns in 45,X females. For a discussion of various neurocognitive applications of the EEG brainwave frequency method along with other EEG techniques (event-related potential [ERP] and probe-ERP), the interested reader should consult a recent chapter by Languis and Wittrock (1986). Advances in brain imaging using techniques such as Positron Emission Tomography (PET), Single Photon Emission Computed Tomography (SPECT), and Magnetic Resonance Imaging (MRI) also may provide unique sources of evidence regarding developmental changes in brain activity that underlie the cognitive and behavioral dysfunctions associated with sex chromosome abnormalities.

## LONGITUDINAL RESEARCH

Bender and Berch (Chapter 1) point out that the use of a prospective, longitudinal approach has clarified greatly the developmental significance of being born with a sex chromosome abnormality. The purpose of this section is to describe some of the major characteristics of longitudinal methodology, point out various rationales for its use, and discuss how a researcher's conception (theory) of historical causation should affect decisions of the timing of observations. Much of this section draws on the literature of developmental psychology (Baltes & Nesselroade, 1979; Schaie & Hertzog, 1982). While most SCA researchers are undoubtedly aware of these principles, presenting them in a more formalized framework may not only be of aid to the novice, but also enhance the manner in which extant longitudinal data sets may be viewed.

According to Baltes and Nesselroade (1979), the term "longitudinal method" actually refers to longitudinal-developmental design orientation instead of a specific method. Moreover, they suggest that the purpose of this orientation is the descriptive and explanatory study of both constancy and change in behavior. In this regard they offer the following definition: "Longitudinal methodology involves repeated time-ordered observation of an individual or individuals with the goal of identifying processes and causes of intraindividual change and of interindividual patterns of intraindividual change in behavioral development" (p. 7). As Baltes and Nesselroade note, this definition incorporates the notion of causation or explanation (determining the mechanisms underlying development) as well as description (identifying the form, sequence, and patterning of behavioral development).

Rationales. Five rationales for longitudinal research have been identified (three descriptive and two explanatory):

(1) Direct identification of intraindividual change (or constancy). This type of change can be quantitative and continuous, that is, a change in the frequency of the same class of behaviors across time. It can consist of the transformation of one behavior to another, as with the hypothesized developmental shift between late childhood and early adolescence from a hyperactive/aggressive personality pattern to a depressed/anxious one in 45,X girls (although the evidence to date for this purported change has only come from cross-sectional studies). Finally, it can also consist of changes in patterns or classes of behaviors considered to be reflective of qualitative or stage-like shifts in some theoretically-based psychological processes, as in

Piaget's hypothesized sequential/hierarchical changes in cognitive systems.

(2) Direct identification of interindividual variability in intraindividual change. This rationale is concerned with the examination of similarities and differences in developmental patterns among different individuals. For example, as pointed out by Bender and Berch (Chapter 1), there are certain behavioral features that occur with increased frequency in SCA subtypes, but not invariably within SCA individuals. Much of this evidence is currently described in terms of individual differences among SCA subjects at different points in their developmental histories. The rationale being addressed here is aimed at determining the degree of homogeneity manifested by different individuals "during the course of intraindividual change." In other words, even more may be gained by examining variations in growth functions or change trajectories in terms of the form, rate and timing of developmental changes.

(3) Analysis of interrelationships among intraindividual changes. This goal involves the use of a multivariate perspective for representing constancy and change in combinations of attributes rather than in any one characteristic per se. Such a holistic approach is necessary for identifying systems changes as well as the progressive differentiation of processes. For example, as is evident in a number of chapters in this volume, there is much to be gained by examining the conjoint development of neurophysiological and endocrine systems, cognitive abilities, and psychosocial adaptation in SCA individuals.

(4) Analysis of determinants of intraindividual change. In order to make causal inferences, one must identify time-ordered factors that precede the phenomena being investigated. Longitudinal analysis of antecedent factors becomes especially critical when the causal process is multidirectional in nature, involves a multivariate patterning of influences, and/or when the antecedent factors produce delayed consequences. The latter two conditions are especially relevant for studying what may be rather complex chains of explanatory factors that contribute to various dysfunctions in the cognitive and psychosocial development of SCA individuals.

(5) Analysis of interindividual variability in intraindividual change. Of concern here is that different causal linkages may be responsible for both similar and different patterns of intraindividual change. For example, that a majority of 47,XXY boys manifest similarities in the development of a behavioral style characterized by passivity and unassertiveness does not preclude the possibility that different causal patterns are operating for different boys. While it is obvious that differences in causal influences may produce interindividual differences in intraindividual change, it should be pointed out that these can be composed not only of variations in the actual content or

substance of such influences, but also of variations in the intensity, timing, and/or patterning of the same basic determining factors (Baltes & Nesselroade, 1979).

A final note regarding analysis of the causal patterns of developmental change concerns the distinction between concurrent and historical explanation. Concurrent refers to causal factors that are proximal to the consequent outcome, while historical refers to agents that bear a distal temporal relationship to their sequelae. While both of these may influence the outcome at any given point in the process of behavioral development, "the developmental orientation to the study of behavior becomes increasingly salient the more 'historical' the phenomenon (in terms of both description and explanation), the longer the causal chain, and the more distant its causal origin" (Baltes & Nesselroade, 1979, p. 36).

Of particular importance to the analysis of historical causation is the notion of causal lag or the so-called sleeper effect. Baltes and Nesselroade (1979) make a distinction between two types of causal lags. With one type, antecedent factors lead inevitably to a certain outcome, albeit through a lengthy causal process. Medical examples they provide include the delayed effects (20 years) of atomic radiation in Japan on the occurrence of leukemia, and slow acting viruses such as kuru. The second type of sleeper effect consists of what Baltes and Nesselroade refer to as a time-lagged multicausality-contingency relationship. Here, while a given causal factor is assumed to have operated at an earlier time in the individual's developmental history, its effect does not become apparent until he/she encounters at least one precipitating event. Typically, examples of this type consist of a presumed genetic predisposition and subsequent ontogenetic factors that may include either physical or psychosocial events or some combination of these.

Obviously, the latter type of sleeper effect is precisely what SCA researchers are interested in studying. Yet as Baltes and Nesselroade point out, the possible occurrence of causal lags necessitates special consideration in designing a longitudinal study. Drawing on the suggestions of previous investigators, they note that good decisions about the timing of observations (including frequency, duration, and spacing) can be made only if there exists a sound theory of the causal-lag process. Since such theories are rare, Baltes and Nesselroade suggest that one may need to vary aspects of the duration and spacing of observations for subgroups of subjects so as to maximize the probability of identifying explanatory mechanisms as well as developmental processes. Of course, while such an approach is less feasible with the relatively small samples available for the longitudinal study of SCA development, the concepts described here should at least be taken into consideration when investigators formulate the observational parameters of their longitudinal measurement plan.

## Practical and Methodological Problems

Based on their own experiences and those of other researchers, Harway, Mednick, and Mednick (1984) have recently summarized the practical difficulties encountered by those who have undertaken the longitudinal approach. These include costs, funding, timeliness and inclusiveness, data storage, staffing, and publication record.

These authors point out that the most frequently cited objection to longitudinal research concerns the cost of collecting data over a protracted time period and supporting research staff over the long term. Yet Harway et al. argue cogently that in comparison to other research methodologies, such as the cross-sectional approach, longitudinal research may actually be more cost effective for obtaining the amount and breadth of data normally acquired. They also point out that because of the general belief that longitudinal research is more costly than other types of methods, funding agencies are reluctant to provide continuing support. Harway et al. note that even if investigators have received government funding for a number of years, their support may be terminated abruptly halfway through their project because the agency feels that the study has gone on for "long enough."

As one might expect, a number of problems emerge in the process of storing the data from a longitudinal project, due to the large volume of information collected and the resulting amount of clerical time needed for processing it. Harway et al. note that while the advent of computers has simplified these tasks, data that have been lost or improperly processed may not be redeemable either by re-collection or reconstruction. Thus they strongly recommend that researchers conduct periodic evaluations using quality control procedures during the data-collection phase of the study.

Another practical difficulty involves changes in staffing. Very few of the staff are likely to remain with a project for its duration. While training new personnel is costly, as Harway et al. point out, periodic staff changes may provide new intellectual blood leading to suggestions for additional assessment techniques and/or alternative data analysis procedures. The latter suggestion is linked to another practical problem: the timeliness and inclusiveness of theories and instrumentation. Harway et al. note that, "Information needed to pursue a hypothesis developed at a later time may not have been collected at the inception of the project because its possible significance was not perceived" (p. 24). To avert such problems, Harway et al. provide a number of suggestions for good planning, only some of which are relevant to SCA research. First, the design should be flexible enough to permit examination of the effects of unanticipated factors. Second, the theoretical conceptions guiding the research should be kept fairly general, allowing for changes over time. Third, data should be recorded in raw form to permit later reanalysis and interpretation in light of theoretical and empirical advances.

A final concern for longitudinal researchers is that of publication. Harway et al. point out that since some of the early investigators involved in longitudinal research were relatively unproductive, the belief remains today that there is a scarcity of published longitudinal findings. However, these authors argue that the "politics of professional recognition" conceals the vast amount of longitudinal research that has in fact been published. That is, they suggest that many longitudinal researchers prefer to publish in journals within their own discipline, because this approach enhances their prestige more than by publishing in an interdisciplinary journal or in a journal from an entirely different discipline.

Harway et al. also provide a discussion of various methodological issues that arise in the design and conduct of longitudinal research. Those to be discussed here include sampling, follow-up, repeated measures, and data collection intervals.

Issues related to sampling are naturally of critical importance to SCA researchers. While Harway et al. do not discuss considerations that bear upon SCA research, such factors will be raised here. As Bender and Berch (Chapter 1) have pointed out, many of the early studies of SCA individuals suffered from various kinds of sampling bias. It was not until the advent of prospective studies involving newborn screening of unselected samples through karyotyping that the generalization of behavioral findings could not be constrained by sampling biases. However, a number of researchers continue to conduct studies with relatively small, fortuitous samples. Since 45,X females typically present with physical anomalies either at birth, during early childhood, or preadolescence, they are less likely to go unidentified than other SCA types. Consequently, studies of selected samples of these females are less likely to be subject to sampling bias. Nevertheless, as Rovet and others in this volume have cogently pointed out, a multicenter collaborative approach may be necessary if future investigations are to be better controlled and more methodologically sound.

Another important problem that arises in longitudinal research is attrition. However, as Harway et al. note, the use of efficient administrative follow-up procedures have been shown to be quite effective in reducing this form of bias. Since longitudinal studies of SCA individuals involve relatively small sample sizes, the problem of attrition can become especially pernicious. Yet in comparison with longitudinal studies of nonspecial populations, other medical and/or psychological factors increase the likelihood of continued participation by SCA subjects. That is, since these individuals are likely to have at least annual checkups with endocrinologists, neurologists, or other specialists in a clinical or hospital setting, SCA behavioral researchers are often able to schedule a testing session for the same day as the medical visit.

The repeated measurement of subjects presents problems of its own. Namely, a testing bias might be operating, where experience with any given test may have an impact on subsequent retestings with the same instrument. The use of an appropriate control group can obviate this problem. Another potential difficulty that can arise is concerned with the selection of data-collection intervals. While this choice is affected in part by factors not under the control of the researcher, Harway et al. offer some considerations for deciding upon an appropriate measurement interval. The most important of these is determining the rate of change of the characteristics being studied.

An additional problem that may arise for SCA researchers is bias due to awareness of the diagnosis of the participants. That is, at least for some kinds of assessment, "experimenter effects" may occur if, for example, the knowledgeable examiner anticipates a lower level of performance on a given test for the SCA individuals as compared to their matched controls. Bender et al. (1983) describe their use of a single-blind procedure to circumvent this problem. With this approach, the examiner is not informed as to whether the subject is in the SCA or control group. However, as Bender et al. note, the obvious physical characteristics comprising the Turner syndrome precludes the use of this procedure with 45,X females.

## STRUCTURAL EQUATION MODELING

A set of procedures that has found increasing application to the analysis and interpretation of data derived from quasi-experimental and nonexperimental research comes under the broad heading of structural equation modeling (SEM). This term refers to a set of techniques that includes causal modeling, path diagrams, regression analysis, and maximum-likelihood estimation (e.g., LISREL) (Biddle & Marlin, 1987). The reader may be interested to learn that the roots of structural equation modeling can be traced to early work in the field of population genetics (viz., to the path-analytic procedure developed by Sewall Wright, [1921, 1931]). Nevertheless, these techniques are considered somewhat controversial today in terms of the appropriateness of their applicability to the field of genetic epidemiology (Cloninger, Rao, Rice, Reich, & Morton, 1983; Karlin, Cameron, & Chakraborty, 1983; Wright, 1983). Furthermore, while structural equation modeling has been applied rather extensively to longitudinal adoption projects in the area of developmental behavioral genetics (e.g., DeFries, Plomin, & LaBuda, 1987), it has not as yet been used for examining the causal structures of developmental changes in SCA individuals. Certainly, the potential utility of such techniques for the study of SCA development makes them worthy of at least a noncomputational overview of their essential components. Interested readers may want to consult a recent introductory textbook for further procedural details (Loehlin, 1987).

One begins by devising a model of "causal" relationships among two or more "independent" (exogenous or predictor) variables and any number of dependent (endogenous or criterion) variables. It is important for the reader to understand that the statistical procedures employed in causal modeling are used primarily for the purpose of testing hypotheses about causal connections, rather than for discovering causal connections (Mulaik, 1987). Anderson (1987) has outlined four types of causal models: recursive, nonrecursive, longitudinal, and latent-variable. In recursive models, the proposed causal links are nonreciprocal; that is, they are presumed to occur in only one direction. Put another way, no two variables are related in such a manner that each depends on the other, nor does any variable feed back upon itself via other variables. In contrast, nonrecursive models include hypothesized causal links that are reciprocal in nature.

Longitudinal or so-called first-order autoregressive models represent variables as causes of themselves over two points in time. In this regard, Hertzog and Nesselroade (1987) have recently provided a critical appraisal of the application of this type of model to longitudinal, repeated-measurements data. Basically, they suggest that since such models are implicitly based on a trait conception of the variables specified, they fail to take into account known subtleties regarding conceptions of stability and change.

The final type of model, latent-variable, involves the postulating of one or more intervening or latent variables as mediating the influence of at least one of the independent variables on the dependent variable. Latent variables are unobserved, statistical abstractions or hypothetical constructs which may be operationalized via multiple indicators that yield a common factor. However, since latent variables cannot be measured directly, precise inferences concerning their substantive meaning may be equivocal (Bentler, 1980). Nevertheless, the chief advantage of latent-variable models is that they are more appropriate for problems of explanation and causal understanding than models based only on so-called manifest (observed or measured) variables. The latter are more useful for problems of description and prediction.

As Bentler (1980) describes it, in latent-variable models, the relations among all constructs and the relations of all constructs to manifest variables are specified in mathematical form. The resulting structural equations (one for each dependent variable) constitute a set of simultaneous, linear regression equations. The model purportedly explains the statistical properties of the manifest variables in terms of the hypothetical latent variables.

The hypothesized relationships within structural equations are frequently expressed in the form of path diagrams, which display them graphically. When used for expressing results, path diagrams usually include numerical information, such as correlation coefficients, standardized partial regression coefficients, and coefficients of determination. The use of least-squares regression procedures permits the investigator to assess how much variance in the dependent variable is accounted for by each explanatory variable (either independent or latent), when the effects of the remaining variables in the set are held constant. Several stages of analysis may be necessary for examining all posited causal relationships. More advanced techniques such as LISREL permit one to determine the degree to which the model fits the data as a whole, in one stage of analysis. If the proposed model does not provide a good fit to the data (by virtue of the statistical test yielding a sufficiently large chi-square value), it is rejected. If the model cannot be rejected statistically, it is considered to constitute a plausible representation of the causal (i.e., process or system) structure (Bentler, 1980).

## Small Sample Research

One reason for the lack of use of SEM techniques with SCA individuals may be the fact that problems arise when applying these techniques to small samples. Moreover, as Tanaka (1987) points out, the problem of adequate sample size is compounded in testing latent-variable structural equation models, because statistical theory is asymptotic in nature, implying that one can draw conclusions confidently from data only as total sample size increases without bound. Unlike univariate statistical models, statistical parameters cannot be adjusted for sample-size differences. Building upon recent statistical developments, Tanaka (1987) has devised an estimation strategy for use with small samples, where the ratio of number of subjects to number of variables is less than adequate. He demonstrates the utility of his procedure with previous data from a study of 50 respondents where the subject-to-variable ratio was approximately 4:1. Since inferential statistical theory has yet to be developed for Tanaka's procedure, he suggests that its most immediate utility may lie in exploratory rather than confirmatory uses of structural equation modeling.

## Application to the Study of SCA Development

One of the ways in which structural equation modeling may be applied to the study of SCA development is that of providing a framework for illustrating putative causal connections between various biochemical, physiological, and behavioral variables. For example, with regard to 47,XXY males, Netley (Chapter 6) has proposed a theory linking fetal hormone levels, cell division rates, hemispheric specialization, cognitive dysfunctions, and psychosocial problems. The specific hypothesized relationships could be expressed graphically in the form of a path diagram. This would not only clarify the nature of the proposed relationships for the reader, but also might yield additional hypotheses that have yet to be tested.

## CONCLUSIONS

Despite the vast array of potential methodological pitfalls that awaits the prospective SCA researcher, those of us who have chosen to participate in this endeavor (and continued with it long enough to be represented in this volume) can attest to the unparalleled experiences (positive as well as negative) of working in this field. We realize that the study of SCA individuals can provide a unique perspective on the confluence of factors contributing to human psychological development. Yet further progress in this area will require increased methodological awareness and sophistication. Unquestionably few, if any, extant SCA studies approximate the ideal investigation as outlined within these pages. In fact it may be next to impossible to attain such an objective even in the near future, if for no other reason than because of the complexities associated with initiating a multicenter (read multination) collaborative project. As Kail (1985) has stated in his discussion of the methodologically ideal study concerning use of response times in cognitive development, "In the interim, the ideal study should remind us of the need to be cautious in claims based on studies that are less than ideal" (p. 272).

## ACKNOWLEDGMENTS

I wish to express my thanks to Bruce G. Bender, William N. Dember, and Carrie Mason-Rogers for their helpful comments on various drafts of the manuscript. I am also grateful to Karen L. Kirkendall for her extensive contributions to all phases of our research concerning spatial information processing in children with Turner syndrome, which formed the basis of our presentation at the AAAS symposium. The writing of this chapter was supported in part by the James and Annette Rosenthal Fund and Grant No. MCJ:000-912-23-0, awarded by the Bureau of Health Care Delivery and Assistance, Division of Maternal and Child Health, Public Health Service, DHHS, and Grant No. 07DD0269/07, awarded by Administration on Developmental Disabilities, OHDS, DHHS.

## REFERENCES

Anderson, J. G. (1987). Structural equation models in the social and behavioral sciences: Model building. Child Development, 58, 49-64.

Anderson, S., Auequier, A., Hauck, W. W., Oakes, D., Vandaele, W., & Weisberg, H. I. (1980). Statistical methods for comparative studies. New York: Wiley.

Baltes, P. B., & Nesselroade, J. R. (1979). History and rationale of longitudinal research. In J. R. Nesselroade & P. B. Baltes (Eds.), Longitudinal research in the study of behavior and development, (pp. 1-39). New York: Academic Press.

Baumeister, A. A. (1967). Problems in comparative studies of mental retardates and normals. American Journal of Mental Deficiency, 71, 869-875.

Bender, B., Fry, E., Pennington, B., Puck, M., Salbenblatt, J., & Robinson, A. (1983). Speech and language development in 41 children with sex chromosome anomalies. Pediatrics, 71, 262-267.

Bentler, P. M. (1980). Multivariate analysis with latent variables: Causal modeling. In M. R. Rosenzweig & L. W. Porter (Eds.), Annual review of psychology. Vol. 31 (pp. 419-456). Palo Alto: Annual Reviews, Inc.

Berch, D. B., Hartmann, L. A., & Bofinger, M. (1986, April). Memorial comparisons in Turner syndrome. Paper presented at the meeting of the Conference on Human Development, Nashville, TN.

Berch, D. B., & Kirkendall, K. L. (1986, May). Spatial information processing in 45,X children. In A. Robinson (Chair), Cognitive and psychosocial dysfunctions associated with sex chromosome abnormalities. Symposium presented at the meeting of the American Association for the Advancement of Science, Philadelphia.

Berch, D. B., & Kirkendall, K. L., Briscoe, G., Dignan, P. St. J., & Smith, K. L. (1985, April). Spatial information processing in children with Turner syndrome. Paper presented at the meeting of the Society for Research in Child Development, Toronto.

Berk, R. A. (1982). Verbal-Performance IQ discrepancy score: A comment on reliability, abnormality, and validity. Journal of Clinical Psychology, 38, 638-641.

Biddle, B. J., & Marlin, M. M. (1987). Causality, confirmation, credulity, and structural equation modeling. Child Development, 58, 4-17.

Bogartz, R. S. (1976). On the meaning of statistical interactions. Journal of Experimental Child Psychology, 22, 178-183.

Bowen, C., Gelabert, T., & Torgesen, J. (1978). Memorization processes involved in performance on the Visual-Sequential Memory subtest of the Illinois Test of Psycholinguistic Abilities. Journal of Educational Psychology, 70, 887-893.

Bradshaw, J. L., Burden, V., & Nettleton, N. C. (1986). Dichotic and dichhaptic techniques. Neuropsychologia, 24, 79-90.

Bradshaw, J. L., & Nettleton, N. C. (1983). Human cerebral asymmetry. Englewood Cliffs, NJ: Prentice-Hall.

Broadbent, D. E. (1954). The role of auditory localization and attention in memory span. Journal of Experimental Psychology, 47, 191-196.

Brunner, R. L., Berch, D. B., & Berry, H. (1987). Pheylketonuria and complex spatial visualization: An analysis of information-processing. Developmental Medicine and Child Neurology, 29, 460-468.

Bryden, M. P., Saxby, L. (1986). Developmental aspects of cerebral lateralization. In J. E. Obrzut and G. W. Hynd (Eds.), Child neuropsychology: Vol. 1: Theory and research, (pp. 73-94). Orlando: Academic Press.

Caruso, R., & Detterman, D. K. (1983). Stimulus encoding by mentally retarded and nonretarded adults. American Journal of Mental Deficiency, 87, 649-655.

Chadwick, O., & Rutter, M. (1983). Neuropsychological assessment. In M. Rutter (Ed.), Development neuropsychiatry, (pp. 181-212). New York: Guilford.

Chapman, L. J., & Chapman, J. P. (1973). Problems in the measurement of cognitive deficit. Psychological Bulletin, 79, 380-385.

Chapman, L. J., & Chapman, J. P. (1974). Alternatives to the design of manipulating a variable to compare retarded and nonretarded subjects. American Journal of Mental Deficiency, 79, 404-411.

Chase, W. G. (1978). Elementary information processes. In W. K. Estes (Ed.), Handbook of learning and cognitive processes: Vol. 5: Human information processing, (pp. 19-90). Hillsdale, NJ: Erlbaum.

Cloninger, C. R., Rao, D. C., Ricke, J., Reich, T., & Morton, N. E. (1983). American Journal of Human Genetics, 35, 733-756.

Cohen, J. (1957). A factor-analytically based rationale for the Wechsler Adult Intelligence Scale. Journal of Consulting Psychology, 21, 451-457.

Cohen, J. (1959). The factorial structure of the WISC at ages 7-6, 10-6, and 13-6. Journal of Consulting Psychology, 23, 285-299.

Das, J. P. (1985). Aspects of digit-span performance: Naming time and order memory. American Journal of Mental Deficiency, 89, 627-634.

DeFries, J. C., Plomin, R., & LaBuda, M. C. (1987). Genetic stability of cognitive development from childhood to adulthood. Developmental Psychology, 23, 4-12.

Dempster, F. N. (1981). Memory span: Sources of individual differences. Psychological Bulletin, 89, 63-100.

Detterman, D. K. (1979). Memory in the mentally retarded. In N. R. Ellis (Ed.), Handbook of mental deficiency, psychological theory and research, (2nd Ed., pp. 727-760). Hillsdale, NJ: Erlbaum.

Donders, F. C. (1969). On the speed of mental processes. Acta Psychologica, 30, 412-431. [Translated from the original by W. G. Koster from Onderzoekingen gedaan in het Physiologisch Laboratorium der Utrechtsche Hoogeschool, 1868, Tweede reeks, II, 92-120.]

Elliot, R. (1970). Simple reaction time: Effects associated with age, preparatory interval, incentive shift, and mode of presentation. Journal of Experimental Child Psychology, 9, 86-104.

Feuerstein, M., Ward, M. M., & LeBaron, S. W. M. (1979). Neuropsychological and neurophysiological assessment of children with learning and behavior problems: A critical appraisal. In B. B. Lahey & A. E. Kazdin (Eds.), Advances in clinical child psychology: Vol. 2 (pp. 241-278). New York: Plenum.

Field, J. G. (1960). Two types of tables for use with Wechsler's intelligence scales. Journal of Clinical Psychology, 16, 3-7.

Garron, D. (1977). Intelligence among persons with Turner's syndrome. Behavior Genetics, 7, 105-127.

Grossman, F. M. (1983a). Percentage of WAIS-R standardization sample obtaining verbal-performance discrepancies. Journal of Consulting and Clinical Psychology, 51, 641-642.

Grossman, F. M. (1983b). Interpreting WISC-R verbal-performance discrepancies: A note for practitioners. Perceptual and Motor Skills, 56, 96-98.

Hahn, W. K. (1987). Cerebral lateralization of function: From infancy through childhood. Psychological Bulletin, 101, 376-392.

Harway, M., Mednick, S. A., & Mednick,. B. (1984). Research strategies: Methodological and practical problems. In S. A. Mednick, M. Harway, & K. M. Finello (Eds.), Handbook of longitudinal research: Vol 1: Birth and childhood cohorts, (pp. 22-30). New York: Praeger.

Hertzog, C., & Nesselroade, J. R. (1987). Beyond autoregressive models: Some implications of the trait-state distinction for the structural modeling of developmental change. Child Development, 58, 93-109.

Hunt, E., & Pellegrino, J. (1985). Using interactive computing to expand intelligence testing: A critique and prospectus. Intelligence, 9, 207-236.

Kail, R. (1985). Interpretation of response time in research on the development of memory and cognition. In C. J. Brainerd & M. Pressley (Eds.), Basic processes in memory development: Progress in cognitive development research, (pp. 249-275). New York: Springer-Verlag.

Kail, R., & Bisanz, J. (1982). Cognitive development: An information-processing perspective. In R. Vasta (Ed.), Strategies and techniques of child study, (pp. 209-243). New York: Academic Press.

Karlin, S., Cameron, E. C., & Chakraborty, R. (1983). Path analysis in genetic epidemiology: A critique. American Journal of Human Genetics, 35, 695-732.

Kaufman, A. S. (1975). Factor analysis of the WISC-R at eleven age levels between 6 1/2 and 16 1/2 years.. Journal of Consulting and Clinical Psychology, 43, 135-147.

Kaufman, A. S. (1976). Verbal-Performance IQ discrepancies on the WISC-R. Journal of Consulting and Clinical Psychology, 44, 739-744.

Kaufman, A. S. (1979). Intelligent testing with the WISC-R. New York: Wiley-Interscience.

Kimura, D. (1961). Cerebral dominance and the perception of verbal stimuli. Canadian Journal of Psychology, 15, 166-171.

Languis, M., & Wittrock, M. C. (1986). Integrating neuropsychological and cognitive research: A perspective for bridging brain-behavior relationships. In J. E. Obrzut & G. W. Hynd (Eds.), Child neurology: Vol. 1: Theory and research, (pp. 209-239). Orlando: Academic Press.

Loftus, G. R. (1978). On interpretation of interactions. Memory and Cognition, 6, 312-319.

Loehlin, J. C. (1987). Latent variable models: An introduction to factor, path and structural analysis. Hillsdale, NJ: Erlbaum.

Milgram, N. A. (1973). Cognition and language in mental retardation: Distinctions and implications. In D. K. Routh (Ed.), The experimental psychology of mental retardation, (pp. 157-230). Chicago: Aldine.

Money, J., & Alexander, D. (1966). Turner syndrome. Further demonstration of the presense of specific congenital deficiencies. Journal of Medical Genetics, 3, 47-48.

Moscovitch, M. (1986). Afferent and efferent models of visual perceptual asymmetries: Theoretical and empirical implications. Neuropsychologia, 24, 91-114.

Mulaik, S. A. (1987). Toward a conception of causality applicable to experimentation and causal modeling. Child Development, 58, 18-32.

Mulder, G. (1983). The information-processing paradigm: Concepts, methods, and limitations. Journal of Child Psychology and Psychiatry, 24, 19-35.

Pachella, R. G. (1974). The interpretation of reaction time in information-processing research. In B. H. Kantowitz (Ed.), Human information processing: Tutorials in performance and cognition, (pp. 41-82). Hillsdale, NJ: Erlbaum.

Pellegrino, J. W., Hunt, E. B., Abate, R., & Farr, S. (1987). A computer-based test battery for the assessment of static and dynamic spatial reasoning abilities. Behavior Research Methods, Instruments, and Computers, 19, 231-236.

Pennington, B. F., Heaton, R. K., Karzmark, P., Pendleton, M. G., Lehman, R., & Shucard, D. W. (1985). The neuropsychological phenotype in Turner syndrome. Cortex, 21, 391-404.

Rovet, J., Netley, C. (1982). Processing deficits in Turner's syndrome. Developmental Psychology, 18, 77-94.

Sanders, A. F. (1980). Stage analysis and response preparation. In G. E. Stelmach and J. Requin (Eds.), Advances in psychology: Vol 1: Tutorials in motor behavior, (pp. 331-354). Amsterdam: North-Holland.

Sattler, J. M. (1982). Assessment of children's intelligence and special abilities, (2nd Ed.). Boston: Allyn & Bacon.

Schaie, K. W., & Hertzog, C. (1982). Longitudinal methods. In B. B. Wolman (Ed.), Handbook of developmental psychology, (pp. 91-115). Englewood Cliffs, NJ: Prentice-Hall.

Segalowitz, S. J. (1986). Validity and reliability of noninvasive lateralization measures. In J. E. Obrzut and G. W. Hynd (Eds.), Child neuropsychology: Vol. 1: Theory and research, (pp.191-208). Orlando: Academic Press.

Shaffer, J. W. (1962). A specific cognitive deficit observed in gonadal aplasia (Turner's syndrome). Journal of Clinical Psychology, 18, 403-406.

Silverstein, A. B. (1981). Reliability and abnormality of test score differences. Journal of Clinical Psychology, 37, 392-394.

Stanovich, K. E. (1978). Information processing in mentally retarded individuals. In N. R. Ellis (Ed.), International review of research in mental retardation, Vol. 9, (pp. 29-60). New York: Academic Press.

Sternberg, S. (1966). High speed scanning in human memory. Science, 153, 652-654.

Sternberg, S. (1969). The discovery of processing stages: Extensions of Donders' method. In W. G. Koster (Ed.), Attention and performance II. Acta Psychologica, 30, 276-315.

Stewart, K. J., & Moely, B. E. (1983). The WISC-R third factor: What does it mean? Journal of Consulting and Clinical Psychology, 51, 940-941.

Tanaka, J. S. (1987). "How big is big enough?": Sample size and goodness of fit in structural equation models with latent variables. Child Development, 58, 134-146.

Taylor, D. A. (1976). Stage analysis of reaction time. Psychological Bulletin, 83, 161-191.

Torgesen, J. (1975). Problems and prospects in the study of learning disabilities. In E. M. Hetherington (Ed.), Review of child development research, Vol. 5, (pp. 385-440). Chicago: University of Chicago Press.

Torgesen, J. (1979). What shall we do with psychological processes? Journal of Learning Disabilities, 12, 16-23.

Waber, D. P. (1979). Neuropsychological aspects of Turner's syndrome. Developmental Medicine and Child Neurology, 21, 58-70.

Wechsler, D. (1974). Manual for the Wechsler Intelligence Scale for Children-Revised. New York: Psychological Corporation.

Wright, S. (1921). Correlation and causation. Journal of Agricultural Research, 20, 557-585.

Wright, S. (1931). Statistical methods in biology. Journal of the American Statistical Association, 26, 155-163.

Wright, S. (1983). On "Path analysis in genetic epidemiology: A critique". American Journal of Human Genetics, 35, 757-768.

*Shelley D. Smith*

# 10 The Contribution of Studies on Sex Chromosome Aneuploidies to the Understanding of Genetic Influences on Behavior

The studies described in this volume have been and will continue to be invaluable in the domains of clinical genetics and developmental psychology. These two fields have much to offer each other, but there have not been many studies that have examined their overlap so thoroughly. Descriptions of clinical syndromes in the genetic literature often concentrated on physical findings and overall IQ range, and rarely mentioned specific disabilities or developmental characteristics. Conversely, investigations of behavioral traits were often done without regard to possible genetic variation within the population. Aside from studies of very common disorders such as Down syndrome and Fragile X, there has been relatively little examination of the developmental characteristics of groups with other, less frequently occurring genetic anomalies. Part of the reason for this lack of research in the rarer syndromes is simply that it is difficult to assemble enough subjects for a comprehensive study. The studies of individuals with sex chromosome aneuploidies (SCA) stand in contrast to the above in that detailed studies of cognition and personality are available and exemplify the international and interdisciplinary collaboration necessary to accomplish them (Stewart & Greene, 1982).

Studies of the development of behavioral characteristics of known genetic syndromes are important to the geneticist in the provision of accurate genetic counseling and to the understanding of how genes can influence behavior. These studies are equally important to the psychologist in understanding normal development and the factors that can disrupt it. Finally, the methods of research developed for the study of SCA, as detailed in these chapters, will provide a guide for similar studies of genetics and behavior.

## MEDICAL GENETICS

A major part of genetic counseling for any condition is to help the family understand both the prognosis and the limits of prognostication. The need for accurate, unbiased information has been especially critical in dealing with SCA because of the highly publicized misinformation, as Bender and Berch detail in Chapter 1. The neonatal screening projects were critical in identifying an unbiased population for prospective, longitudinal follow-up, and the careful methods of the investigators took into account the possible pitfalls of stigmatization and "self-fulfilling prophecy" (Robinson et al., 1982). As noted by Bender, Linden, and Robinson (Chapter 2), by Rovet (Chapter 3), and by Ratcliffe, Jenkins, and Teague (Chapter 8), one of the most critical times when accurate counseling is needed is when an SCA is detected prenatally.

Discussion of prognosis with the family should include "anticipatory guidance" (Bender, Linden, & Robinson, Chapter 2). Possible problems that could arise and the appropriate intervention should be discussed. The demonstrated variability within children with the same SCA means that counseling must emphasize what could be rather than what will be. The research also points out that environmental and other genetic factors are important in determining the ultimate phenotype. The evaluation of a child with an SCA may need to include assessments of family functioning, for example, to identify potential problems and plan effective intervention.

Information is still lacking regarding the long-term outcomes for adults with SCA. Some of the children identified in the longitudinal studies are now reaching adulthood, and it will be very interesting to determine the behavioral characteristics that (1) are maintained with development, (2) disappear with eventual maturation, (3) emerge during later developmental phases, and (4) are modified by environmental factors or therapies. For example, Bender et al. question whether the risk of psychosocial adjustment problems found in adolescents with neurocognitive deficits will persist into adulthood. McCauley wonders whether the immaturity seen in girls with Turner Syndrome (TS) is partially a reaction to being perceived as younger by those around them, and proposes that the immaturity might resolve with attainment of an adult appearance. Understanding the natural history of SCA will not only help in providing accurate prognostic information, but also will be important in determining underlying mechanisms for findings noted earlier in life and in assessing long-term results of medical and/or psychological interventions.

## BEHAVIOR GENETICS

Continued study of SCA individuals will help answer some specific questions of greater interest to the field of behavior genetics. As Rovet states in Chapter 3, TS was the first syndrome to demonstrate "a potential link between genetic, sex-related, and cognitive characteristics" (p. 38). Questions that will be addressed through study of SCA include: How are the genetic effects on the behavioral phenotype mediated? Does the overall amount of chromatin affect brain development, or is the dosage of specific genes the important factor? If the effects are hormonally mediated, how is this accomplished, and how is it related to extra or missing chromatin? Is the central nervous system affected through alterations in cellular differentiation, neuronal migration, or specialization of function? How is a specific ability, such as mental rotation, targeted?

Information from other genetic studies can be brought to bear on some of these questions. For example, it appears that the phenotype of Down syndrome is largely due to a specific region of chromosome 21 (band q22) rather than the amount of extra chromatin (see Van Keuren et al., 1986 for a brief literature review); by analogy, the effects of SCA may be related more to specific genes that are extra or missing rather than to the overall amount of chromatin. Contrary to this hypothesis, the 47,XXY and 47,XYY conditions are cognitively similar albeit chromosomally different. Perhaps the relevant genetic influences are actually quite similar in these two syndromes; in effect, the cognitive phenotype might be mediated by genes that are in the region of homology between the X and Y chromosomes.

Indeed, SCA individuals exemplify a number of important genetic principles, not restricted to the area of behavior genetics. Genetic conditions are rarely discontinuous; rather, one usually sees a range of severity. Initial descriptions of a condition may be biased toward the most severely affected, but subsequently, milder forms often are described as the phenotypic pattern is recognized and studies are expanded.

Some of the variation between individuals may be due to underlying heterogeneity; that is, the same phenotype may be caused by different genes ("genocopy") or there may be a nongenetic cause that mimics the genetic syndrome ("phenocopy"). Studies within families, as well as linkage analysis, can help assure that different individuals actually have the same syndrome so that the full phenotypic range can be appreciated. For example, neurofibromatosis is an autosomal dominant condition that is highly variable, with symptoms ranging from a few cafe-au-lait spots to multiple tumors of the skin and organs (neurofibromas). Only about 8% of people with neurofibromatosis are mentally retarded (despite earlier estimates as high as 70%), but learning disabilities are seen in about 30% (Riccardi & Eichner, 1986). The phenotypic variation is seen within family members, who would be carrying the same gene, and linkage studies suggest that all families with neurofibromatosis have the same gene (Barker et al., 1987).

Much of the phenotypic variations in genetic conditions must be due to modification by other genes and environment. As stated by Berch in Chapter 9, "chromosomal anomalies have complex and variegated effects on developing physiological and biochemical systems. Moreover, a range of environmental variables may either attenuate or potentiate the influence of biological determinants on behavioral development" (p. 185). This clearly applies to other genetic conditions with other modes of inheritance as well. Autosomal dominant conditions may be more subject to such influences, as reflected in the variations in expressivity (the possible phenotypic characteristics) and penetrance (the percent of individuals with a given genotype that manifest the phenotype), but autosomal recessive conditions may also show variation in these parameters.

The range of the phenotype in a given karyotype or genotype and the influence of other factors may be leading to a refinement of the definition of the term "multifactorial." It was assumed that many common traits in humans with small but significant genetic influence (e.g., recurrence risks around 2-5% in first degree relatives) were due to the additive effect of many genes ("polygenic inheritance") along with environmental effects. However, it appears that in some of these traits, such as cleft lip and palate (Eiberg, Bixler, Nielsen, Conneally, & Mohr, 1987; Marazita, Spence, & Melnick, 1986), the bulk of the liability is due to autosomal recessive genes at only a few loci, with genetic and nongenetic modifiers affecting penetrance and expressivity. Similarly, club foot appears to be due to a major dominant gene along with a minor multifactorial contribution (Wang, Palmer, & Chung, 1988). The improved ability to analyze statistically such a "mixed model" (Lalouel, Rao, Morton, & Elston, 1983) should help determine whether specific behavioral traits are influenced by a specific gene or genes, and the genetic technique of gene localization can help identify the individual gene or genes involved (Housman, Smith, & Pauls, 1985).

Thus, the phenotypic variability seen in SCA individuals is paralleled by many of the single gene disorders. In this light, the variability in the SCA is quite understandable. If a single gene that is identical in all individuals can show wide variation in expression, it is not surprising that chromosomal disorders, which involve homologous loci but not identical alleles, will show variation from person to person. Yet, consistent patterns emerge.

The existence of a major genetic factor influencing a given behavior must be understood in the context of "reaction range" or "genomic repertoire," as described by Bender and Berch (Chapter 1). Within limits, this very important concept of susceptibility should guide research concerning both the detection of phenotypic patterns in a syndrome and the design of intervention strategies. As Theilgaard (Chapter 7) describes the psychological problems of 47,XXY and 47,XYY men, "it is not a question of grave brain damage, but rather of subtle disturbances that may lead to reduced tolerance of psychological and environmental strain" (p. 153) and "the problems are generally not of such magnitude that society needs to take special precautions. [....] The concept of a population at risk should be viewed with due regard to probability and multideterminism" (p. 158). For example, both karyotype and institutionalization or noninstitutionalization affect the IQ range of children with Down syndrome (Centerwall & Centerwall, 1960). Similarly, both genotype and smoking behavior determine susceptibility to lung disease in alpha-1-antitrypsin deficiency (Gadek & Crystal, 1982). In the same manner, the interaction of phenotype, genotype, and environment must be taken into account in assessing the effects of karyotype and short stature on personality in TS individuals.

Just as there is variation within a population with the same genetic etiology, there also seem to be phenotypic similarities between groups. One of the primary reasons for studying behavioral characteristics in populations with different genetic conditions was the hypothesis that etiologically distinct conditions would be phenotypically distinct, and that the study of etiologically distinct groups would uncover different underlying mechanisms. This is analogous to the subtyping literature in the field of reading disability, where it was hypothesized that different phenotypic categories could be described that would reflect different underlying mechanisms. However, there is some evidence that etiologically distinct conditions may not be phenotypically distinct, at least at current levels of measurement of cognitive traits. As Berch notes (Chapter 9), known etiology does not imply a consistent phenotype, given the considerable variation within SCA conditions. As an example, Rovet finds that there is not a homogeneous cognitive profile in TS, with the exception of disability in mental rotation. Conversely, there is much similarity between the SCA karyotypic groups, particularly between the 47,XXY and 47,XYY boys. As Bender et al. (Chapter 2) point out, both groups show tendencies for language and reading problems, and Theilgaard (Chapter 7) reports similar cognitive and neuropsychological findings for both groups. It is of interest to note that evidence exists for phenotypic similarity between individuals with specific reading disability who presumably have different etiologies (Decker & Bender, 1988; Pennington, Lefly, Van Orden, Bookman, & Smith, 1987). Those studies have suggested that the genetic component to reading disability, whether or not it is related to a single gene defect, acts upon a particular step in the reading process, that of phonemic awareness. It will be particularly fascinating to determine whether this same deficit is also evident in the language-based dyslexia of 47,XXY boys as described by Bender, Salbenblatt, and Robinson (1986).

Thus, although the genetic constitution creates identifiable risks, these appear to be neither absolute nor distinctive in SCA individuals, and may operate similarly in other genetic conditions. Discovery of the underlying mechanisms in these disorders should tell us how one particular genetic entity can have such variable effects, and conversely, how different genetic conditions can look so similar. Exploring the ways in which developmental pathways branch and converge, and attempting to identify the junctures at which genes and environment affect them, promises to be exciting for both geneticists and psychologists.

## NORMAL DEVELOPMENT AND MECHANISMS OF DISRUPTION

The study of an "abnormal" state is often undertaken with the hope that the underlying problem will be described and will lead to hypotheses about the pathogenic mechanism. From there, one gets information not only about how to treat the abnormality, but learns about the normal process as well, and further hypotheses may be generated regarding other possible disruptions of that process. Studying behavioral characteristics of SCA individuals can be seen as leading in these directions, even though the behavioral problems themselves are typically not severe.

Take for example Nyborg's (Chapter 5) description of the development of a particular spatio-perceptual ability in TS women as compared to girls with normal karyotypes. The performance deficits exhibited by the TS women in this study were shown to result from persistence in the use of comparatively immature strategies. He proposes that these cognitive differences were related to differences in hormone levels between groups. It would follow then that hormone treatment might alleviate the spatial disability. Nyborg suggests further that modulation of autosomal genes and/or effects on neurotransmitters may constitute the mechanism of action of the hormones, particularly estradiol, influencing the efficiency of cross-modal matching of spatial information. This type of regulation of autosomal genes by sexually-determined factors is termed "sex-limited," as opposed to "sex-linked," in which the variation is influenced by genes on the X or Y chromosomes. Nyborg speculates parenthetically that this mechanism could be responsible for a number of sexually dimorphic traits.

Similarly, Theilgaard (Chapter 7) examines the psychological characteristics of men with 47,XXY or 47,XYY karyotypes and finds evidence of emotional immaturity and reduced inhibition in both groups, as well as incomplete hemispheric lateralization of function, reflecting generalized developmental lag. However, the development of gender identity, sexual behavior, and expression of aggressivity differs for these groups. Differences in testosterone could be partially responsible, but environmental factors, including the reaction of the individual and others to phenotypic stigmata, could also be important in determining the gender identity.

McCauley (Chapter 4) similarly finds immaturity and decreased libido in TS women. Furthermore, she notes the difficulty in determining whether this is attributable to genetic/hormonal factors or an emotional reaction to short stature, physical stigmata, and/or infertility, including self-perception as well as reaction to the perceptions of others. Control groups with short stature or infertility due to conditions other than TS might help resolve this, but the etiology of the short stature or infertility in the controls would have to be specified, since there could be different genetic/hormonal causes with their own effects on behavior.

Just as the psychological characteristics could be secondary reactions to the physical phenotype, the perceptual problems in TS could secondarily produce immature and less socially adept behavior. McCauley hypothesizes that poor interpretation of facial affect, related to deficits in processing in the right cerebral hemisphere, could adversely affect social competence. In an important distinction between primary and secondary phenotypic characteristics, McCauley points out that the behavioral difficulties commonly seen in TS individuals, rather than being inherent parts of the syndrome, could reflect reactions to the problems presented by their karyotype. If features such as immaturity and poor social competence are not integral to the syndrome, therapy should be addressed specifically to those problems.

Ratcliffe, Jenkins, and Teague (Chapter 8) suggest that the characteristics of withdrawal (underreaction), learning problems, and tall stature seen in boys with the 47,XYY karyotype are interrelated and in turn may produce increased frustration and temper tantrums. To further illustrate the bidirectionality of phenotypic and environmental influences, the authors note that the increased family pathology could be a result of the personality problems in these boys, as well as a cause.

Similarly, Netley (Chapter 6) hypothesizes that personality characteristics in 47,XXY males is secondary to cognitive problems. Specifically, he provides evidence that verbal disabilities, due to abnormalities in left hemisphere specialization for language, result in more passivity, social withdrawal, and inactivity. This was observed to change as the boys matured, possibly due to changes in testosterone levels which appeared to interact with the hemispheric organization. Netley also proposes that hemispheric specialization is related to the mitotic rate of the cells of the CNS in development, which would be inversely related to the amount of chromatin. Delayed mitotic rate, caused by the increased amount of chromatin in the 47,XXY cells, would favor right hemisphere specialization. In contrast, the decreased amount of chromatin in 45,X cells would result in faster cell division and favor left hemisphere specialization, with the consequent deficits in right hemisphere (spatial) abilities.

Finally, Rovet (Chapter 3) notes that her own work in addition to that of others has identified impaired spatial abilities and parietal lobe dysfunction in TS females. In examining the hemispheric lateralization of function, Rovet finds weaker left hemisphere asymmetry for verbal tasks and more involvement of the left hemisphere for spatial (normally "right hemisphere") tasks. She discusses three possible mechanisms: the influence of amount of chromatin on cell growth, the effect of hormones on CNS development, and inhibition of the right hemisphere by the left hemisphere (Kinsbourne, 1982).

## HORMONES AND HEMISPHERES: A SYNTHESIS

A common theme emerging from several of these studies is that the cognitive and personality characteristics of SCA individuals may be due to effects on cerebral development, specifically on hemispheric specialization of function. In addition, several authors propose that the behavioral characteristics are mediated by hormones, either pre-

or postnatally. Hypotheses relating hemispheric specialization and hormone levels have also been put forth by other groups studying cerebral lateralization and/or dyslexia. Netley notes that Geschwind, Behan, and Galaburda (Geschwind & Behan, 1982; Geschwind & Galaburda, 1987) have developed a hypothesis relating increased left-handedness, immune disorders, and a male preponderance in dyslexics to the effects of prenatal testosterone on the developing brain and thymus. He points out further that testosterone is thought to be low in 47,XXY babies perinatally, which would seem to argue against the idea that the left hemisphere dysfunction and dyslexia seen in these boys is due to increased prenatal testosterone levels. However, it is possible that the critical period for the testosterone effect occurs earlier in gestation. The Geschwind/Galaburda hypothesis is still being actively tested and refined by many researchers studying dyslexia (e.g., Kinsbourne, 1986; Pennington, Smith, Kimberling, Green, & Haith, 1987). In a similar vein, Gordon and colleagues have been studying the effects of postnatal hormonal fluctuations on the performance of "right hemisphere" and "left hemisphere" tasks (Gordon, Corbin, & Lee, 1986) and have also related cerebral lateralization of function to dyslexia (Harness, Epstein, & Gordon, 1984). It may be that a combination of information from dyslexia research and SCA research will result in the elucidation of the effects of hormones on brain development from early gestation to adulthood.

## RESEARCH METHODOLOGY

Although similarities can be observed across the wide range of studies described in this volume, as well as with other studies in the literature, differences emerge with regard to the specific deficits exhibited by various SCA groups along with the possible underlying mechanisms. Rovet (Chapter 3) and McCauley (Chapter 4) both call attention to methodological problems such as small sample sizes, differences in definition and ascertainment, and the variety of measures used in different studies. Both Rovet (Chapter 3) and Theilgaard (Chapter 7) emphasize that studies should be hypothesis-driven. In Chapter 9, Berch combines these and other observations to make detailed recommendations for studies of phenotype and possible mechanisms in SCA that are applicable for any good research. Several of the recommendations are particularly important for studies combining genetics and behavioral traits.

In both genetics and psychology, the methods of ascertainment of subjects is very important. The way subjects are found affects the ability to do certain genetic analyses, and the criteria for selection affects the homogeneity of the population within and between studies. The sample size and inclusion of appropriate controls affect the validity of the conclusions about the distinctiveness of the phenotype.

The measures used in defining the phenotype are also critical, whether they are used to contrast the phenotypic characteristics of different groups or to define specific groups for genetic analyses. For example, Berch emphasizes that one must be sure what the tests measure, and that they are reliable and well-standardized. Tests of lateralization of cerebral function, which are so critical to the hypotheses relating genotype to phenotype, have been particularly controversial. It is possible that the variation in phenotypes seen in populations that are genotypically or karyotypically the same is due in part to the level of measurement of the phenotype. Measurement of a factor closer to the "gene product" may produce more consistent results.

Limitations of the variables for statistical analysis must also be recognized, and the researcher must be careful not to overinterpret statistical results. A factor or a cluster is a statistical phenomenon and may or may not reflect an underlying biological reality, and thus may not be an appropriate variable for genetic studies. Berch echoes the other authors in asserting that statistical analyses must be used for testing hypotheses instead of attempting to establish causation.

## CONCLUSION

The studies of SCA individuals described in this volume have contributed very valuable information to the fields of developmental psychology, behavioral genetics and clinical genetics. The unanswered questions will be even more valuable in stimulating research that combines these areas with neurology, neuropsychology, and endocrinology, among others. Taken together, this body of work will lead to improved understanding of cognitive and behavioral development in general. Pennington has stated that "the study of genetically influenced LDs may provide for developmental neuropsychology what the study of acquired lesions has provided for adult neuropsychology" (Pennington & Smith, 1988, p. 822). In the same way, the study of variations in SCA can demonstrate the range and mechanisms of genetic effects on cognition and behavior, extending into the normal range. Elucidation of these mechanisms will almost surely be applicable to other disorders as well as to normal variations in these traits.

## REFERENCES

Barker, D., Wright, E., Nguyen, K., Cannon, L., Fain, P., Goldgar, D., Bishop, D. T., Carey, J., Baty, B., Kivlin, J., Willard, H., Waye, J. S., Greig, G., Leinwand, L., Nakamura, Y., O'Connell, P., Leppert, M., Lalouel, J. M., White, R., & Skolnick, M. (1987). Gene for Von

Recklinghausen neurofibromatosis is in the pericentromeric region of chromosome 17. Science, 236, 1100-1102.

Bender, B., Puck, M., Salbenblatt, J., & Robinson, A. (1986). Dyslexia in 47,XXY boys identified at birth. Behavior Genetics, 16, 343-354.

Centerwall, S. A., & Centerwall, W. R. (1960). A study of children with mongolism reared in the home compared to those reared away from home. Pediatrics, 25, 678-685.

Decker, S. N., & Bender, G. G. (1988). Converging evidence for multiple genetic forms of reading disability. Brain and Language, 33, 197-215.

Eiberg, H., Bixler, D., Nielsen, L. S., Conneally, P. M., & Mohr, J. (1987). Suggestion of linkage of a major locus for nonsyndromic orofacial cleft with F13A and tentative assignment to chromosome 6. Clinical Genetics, 32, 129-132.

Gadek, J. E., & Crystal, R. G. (1982). Alpha-1 antitrypsin deficiency. In J. B. Stanbury, J. B. Wyngaarden, D. S. Fredrickson, J. L. Goldstein, & M. S. Brown (Eds.), The metabolic basis of inherited disease (pp. 1450-1467). New York: McGraw-Hill.

Geschwind, N., & Behan, P. (1982). Left-handedness: Association with immune disease, migraine, and developmental learning disorder. Proceedings of the National Academy of Science USA, 799, 5097-5100.

Geschwind, N., & Galaburda, A. M. (1987). Cerebral lateralization: Biological mechanism, associations and pathology. Cambridge: The MIT Press.

Gordon, H. W., Corbin, E. D., & Lee, P. A. (1986). Changes in specialized cognitive function following changes in hormone levels. Cortex, 22, 399-415.

Harness, B. Z., Epstein, R., & Gordon, H. W. (1984). Cognitive profile of children referred to a clinic for reading disabilities. Journal of Learning Disabilities, 17, 346-352.

Housman, D., Smith, S. D., & Pauls, D. (1985). Applications of recombinant DNA to neurogenetic disorders. In D. B. Gray & J.F. Kavanagh (Eds.), Biobehavioral measures of dyslexia (pp. 155-162). Parkton, MD: York Press.

Kinsbourne, M. (1982). Hemispheric specialization and the growth of human understanding. American Psychologist, 37, 411-420.

Kinsbourne, M. (1986). Relationships between non-right-handedness and diseases of the immune systems. Journal of Clinical and Experimental Neuropsychology, 71, 602.

Lalouel, J. M., Rao, D. C., Morton, N. E., & Elston, R. C. (1983). A unified model for complex segregation analysis. American Journal of Human Genetics, 35, 816-826.

Marazita, M. L., Spence, M. A., & Melnick, M. (1986). Major gene determination of liability in cleft lip with or without cleft palate: A multiracial view. Journal of Craniofacial Genetics and Developmental Biology Supplement, 2, 89-97.

Pennington, B. F., Lefly, D. L., Van Orden, G. C., Bookman, M.O., & Smith, S. D. (1987). Is phonology bypassed in normal or dyslexic development? Annals of Dyslexia, 37, 62-89.

Pennington, B. F., & Smith, S. D. (1988). Genetic influences on learning disabilities: An update. Journal of Consulting and Clinical Psychology, 56, 817-823.

Pennington, B. F., Smith, S. D., Kimberling, W. J., Green, P., Haith, M. M. (1987). Left handedness and immune disorders in familial dyslexics: A test of Geschwind's hypothesis. Archives of Neurology, 44, 634-639.

Riccardi, V. M., & Eichner, J. E. (1986). Neurofibromatosis: Phenotype, natural history, and pathogenesis. Baltimore: The Johns Hopkins Press.

Robinson, A., Bender, B., Borelli, J., Puck, M., Salbenblatt, J., Webber, M. L. (1982). Children with sex chromosome aneuploidy: Follow-up studies. Birth Defects: Original Article Series, 18. New York: Alan R. Liss, Inc.

Stewart, D. A., & Greene, S. C. (Eds.) (1982). Children with sex chromosome aneuploidy: Follow-up studies. Birth Defects: Original Article Series, 18. New York: Alan R. Liss, Inc.

Van Keuren, M. L., Watkins, P. C., Drabkin, H. A., Jabs, E. W., Gusella, J. R., & Patterson, D. (1986). Regional localization of DNA sequences on chromosome 21 using somatic cell hybrids. American Journal of Human Genetics, 38, 793-804.

Wang, J., Palmer, R. M., & Chung, C. S. (1988). The role of a major gene in clubfoot. American Journal of Human Genetics, 42, 772-776.

# ABOUT THE CONTRIBUTORS

BRUCE G. BENDER is a Pediatric Neuropsychologist at the National Jewish Center for Immunology and Respiratory Medicine and Associate Professor of Psychiatry at the University of Colorado School of Medicine, Denver, Colorado. He has studied extensively children with sex chromosome abnormalities and children with asthma. He received his B.S. from North Central College and his Ph.D. from the University of Wisconsin.

DANIEL B. BERCH is an Associate Professor of psychology at the University of Cincinnati and Research Coordinator at the University Affiliated Cincinnati Center for Developmental Disorders. His primary research interest is in cognitive development, specializing in the information-processing approach for studying the development of spatial abilities. He obtained his B.A. from the University of Michigan and his Ph.D. from the University of New Mexico.

JENNIFER JENKINS is a Lecturer in psychology at the University of Stirling in Stirling, United Kingdom. She is a clinical psychologist who has conducted research in child psychopathology, family therapy, and various developmental concerns in both well and sick children. She received her B.A. from Sussex University and her Ph.D. from London University.

MARY LINDEN is a Genetic Associate in the Department of Pediatrics at the National Jewish Center for Immunology and Respiratory Medicine in Denver, Colorado. She is a genetic counselor and for the past three years has coordinated the Denver study of children with sex chromosome abnormalities under the direction of Arthur Robinson, M.D. She obtained her B.S. from the University of Wisconsin and her M.S. from the University of Colorado.

ELIZABETH MCCAULEY is an Associate Professor of Psychiatry and Behavioral Sciences at the University of Washington School of Medicine and Children's Hospital and Medical Center in Seattle, Washington. Her research areas include child clinical psychology and psychoendocrinology. She received her B.A. from the University of Wisconsin and her Ph.D. from the State University of New York at Buffalo.

CHARLES NETLEY is a Professor in the Department of Behavioral Science at the University of Toronto. His research has spanned numerous topics in developmental psychology and neuropsychology, including the study of cognitive and personality development in various X aneuploid groups. He obtained his B.A. from Queen's University and his Ph.D. from the University of London.

HELMUTH NYBORG is an Associate Professor at the Institute of Psychology, University of Aarhus, Aarhus, Denmark. His major field of interest is psychoneuroendocrinology, specializing in studying the effects of sex hormones on spatial ability in both normal development and in individuals with hormonal imbalances and sex chromosome abnormalities.

SHIRLEY G. RATCLIFFE is a Pediatrician and Honorary Consultant in the Human Genetics Unit of the Medical Research Council at Western General Hospital, Edinburgh, Scotland. She has studied intensively the growth and development of children with sex chromosome abnormalities. She has obtained her B.S., M.B., and F.R.C.P. degrees.

ARTHUR ROBINSON is a Senior Staff Member at the National Jewish Center for Immunology and Respiratory Medicine, Denver, Colorado, and Professor Emeritus at the University of Colorado School of Medicine. Both a pediatrician and a geneticist, his publications in genetics, and in particular on the subject of sex chromosome abnormalities, have spanned three decades. He received his B.A. from Columbia University and his M.D. from the University of Chicago.

JOANNE F. ROVET is an Assistant Professor in the Departments of Psychology and Pediatrics, Division of Endocrinology at the Hospital for Sick Children in Toronto, Canada. In addition to the study of individuals with Turner syndrome and those with other types of sex chromosome abnormalities, she has conducted research concerning the neuropsychology of pediatric endocrine disorders. She obtained her Ph.D. from the University of Toronto.

SHELLEY D. SMITH is an Associate Professor at Boys Town National Institute for Communication Disorders in Children in Omaha, Nebraska, and Coordinator of Clinical and Genetic Services in the Department of Otolaryngology and Human Communications. Her research interests include genetic influences on speech and language disorders, with a specialization in reading disabilities. She received her B.A. from Grinnell College and her Ph.D. from Indiana University.

PETER TEAGUE is a Research Officer in the Department of Pattern Recognition, Clinical and Population Cytogenetics Unit of Western General Hospital in Edinburgh, Scotland. He has worked with Dr. Shirley Ratcliffe on the longitudinal study of children with sex chromosome abnormalities. He received his B.S. from the University of Bristol and M.Sc. from the University of Birmingham.


ALICE THEILGAARD is an Associate Professor of Medical Psychology and Chief Psychologist at the University of Copenhagen Clinic of Psychiatry, Rigshospital, in Copenhagen, Denmark.  Her research interests are in the areas of psychopharmacology, sex chromosome abnormalities, and clinical psychology.  She has a D.M.Sc. degree.

# INDEX